D0908424

System-on-a-Chip: Design and Test

For a listing of related titles from *Artech House,*
turn to the back of this book.

System-on-a-Chip: Design and Test

Rochit Rajsuman

Artech House
Boston • London
www.artechhouse.com

Library of Congress Cataloging-in-Publication Data
Rajsuman, Rochit.
    System-on-a-chip : design and test / Rochit Rajsuman.
        p.   cm. — (Artech House signal processing library)
    Includes bibliographical references and index.
    ISBN 1-58053-107-5 (alk. paper)
        1. Embedded computer systems—Design and construction.   2. Embedded
    computer systems—Testing.   3. Application specific integrated circuits—Design
    and construction.   I. Title.   II. Series.
    TK7895.E42 R37   2000
    621.39'5—dc21   00-030613
                                                                      CIP

British Library Cataloguing in Publication Data
Rajsuman, Rochit.
    System-on-a-chip : design and test. — (Artech House signal processing library)
    1. Application specific integrated circuits — Design and construction
    I. Title
    621.3'95

    ISBN 1-58053-107-5

Cover design by Gary Ragaglia

© 2000 Advantest America R&D Center, Inc.
3201 Scott Boulevard
Santa Clara, CA 95054

International Standard Book Number: 1-58053-107-5
Library of Congress Catalog Card Number: 00-030613

10 9 8 7 6 5 4 3 2 1

Contents

Preface

This project started as an interim report. The purpose was to communicate to various groups within Advantest about the main issues for system-on-a-chip (SoC) design and testing and the common industrial practices. Over one year's time, a number of people contributed in various capacities to complete this report.

During this period, I also participated in the Virtual Socket Interface (VSI) Alliance's effort to develop various specification documents related to SoC design and testing and in the IEEE P1500 working group's effort to develop a standard for core testing. As a result of this participation, I noticed that SoC information is widely scattered and many misconceptions are spread throughout the community, from misnamed terms to complete conceptual misunderstanding. It was obvious that our interim report would be quite useful for the community as a general publication.

With that thought, I contacted Artech House. The editorial staff at Artech House had already been hearing and reading a lot about system-on-a-chip and was very excited about this project. Considering the rapid technology changes, a four-month schedule was prepared and I set out to prepare the manuscript before the end of 1999. Although I had the baseline material in the form of an interim report, simple editing was not enough. Besides the removal of some sections from the report, many sections and even chapters required a complete overhaul and new write-ups. Similarly, a couple of new chapters were needed. Because of the very aggressive schedule and other internal projects, at times it felt very tedious and tiring. This may have resulted in incomplete discussions in a few sections. I was able to fix descriptions in some sections based on feedback from my colleagues at ARD and from Artech reviewers, but readers may find a few more holes in the text.

The objective of this book is to provide an overview on the present state of design and testing technology for SoC. I have attempted to capture the basic issues regarding SoC design and testing. General VLSI design and testing discussions are intentionally avoided and items described are specific to SoC. SoC is in its early stages and so by no means is the knowledge captured in this book complete. The book is organized into two self-contained parts: (1) design and (2) testing.

As part of the introduction to Part I: Design, the background of SoC and definitions of associated terms are given. The introduction also contains a discussion of SoC design difficulties. Hardware–software codesign, design reuse, and cores are the essential components of SoC; hence, in Chapter 2, these topics are discussed, from product definition (specifications) to deliverable requirements and system integration points of view. Some of these methods are already in use by a few companies, while others are under evaluation by other companies and standards organizations. For design reuse, a strict set of RTL rules and guidelines is necessary. Appendix A includes reference guidelines for RTL coding as well as Lint-based checks for the violations of these rules.

Whereas Chapter 2 is limited to digital logic cores, Chapter 3 describes the advantages and issues associated with using large embedded memories on chips and the design of memory cores using memory compilers. Chapter 3 also provides the specifications of some commonly used analog/mixed-signal cores such as DAC, ADC, and PLLs. Chapter 4 covers design validation at individual cores as well as at the SoC level. This chapter also provides guidelines to develop testbenches at cores and SoC levels. Part I concludes with Chapter 5, which gives examples of cores, core connectivity, and SoC.

As part of the introduction to Part II, a discussion on testing difficulties is given. One major component of SoC is digital logic cores; hence, in Chapter 6, test methodologies for embedded digital logic cores are described. Similar to the design methods for digital logic cores, some of the test methods are already in use by a few companies, while others are under evaluation by other companies and standards organizations. Chapter 6 also provides the test methods for microprocessor and microcontroller cores. These cores can be viewed as digital logic cores, however—because of their architecture and functionality—these cores are the brains of SoC. Subsequently, few items beyond the general logic cores are specific to microprocessor/microcontroller cores. These items are also described in Chapter 6.

In addition to logic cores, large memory blocks are another major component of SoC. Chapter 7 discusses the testing of embedded memories. Testing of embedded analog and mixed-signal circuits is discussed in Chapter 8.

Iddq testing has continuously drawn attention. Besides the discussion on technology-related issues, Iddq testing on SoC has some other unique issues. These issues are discussed in Chapter 9 with design-for-Iddqability and vector generation methods.

A number of other topics that are important for SoC testing are related to its manufacturing environment and production testing of SoC. These items include issues such as at-speed testing, test logistics on multiple testers, and general issues of the production line such as material handling, speed binning, and production flow. Discussion on these topics takes place in Chapter 10. Finally, concluding remarks are given in Chapter 11.

Acknowledgment

First of all, I want to express my thanks to the editorial staff at Artech House for their prompt response, enthusiasm, energetic work, and wonderful treatment. My special thanks are due to Mark Walsh, Barbara Lovenvirth, Jessica McBride, Tina Kolb, Bridget Maddalena, Sean Flannagan, and Lynda Fishbourne. I am also thankful to Artech's reviewers for reading the draft and providing very valuable comments.

Needless to say, I am thankful to the many people at ARD who helped me in one way or another with this work. Without continuous support and encouragement from Shigeru Sugamori, Hiro Yamoto, and Robert Sauer, this book would not have materialized. I specifically want to express my thanks to Robert Sauer for the generous amounts of time he spent reviewing chapter drafts during evenings and weekends and giving me feedback. This help was invaluable in identifying many mistakes and omissions. His feedback together with Artech's reviewers helped me resolve many deficiencies in the text.

I also acknowledge and express my thanks to the design and test community in general for their work, without which no book can be written. Specifically, I want to acknowledge the VSI Alliance for developing various specification documents for SoC design and testing. The ongoing work by the IEEE P1500 Working Group as well as publications by the IEEE and Computer Society Press are gratefully acknowledged. I am also thankful to the IEEE for their permission to use numerous diagrams from various papers.

Part I:
Design

1

Introduction

In the mid-1990s, ASIC technology evolved from a chip-set philosophy to an embedded-cores–based system-on-a-chip (SoC) concept. In simple terms, we define an SoC as *an IC, designed by stitching together multiple stand-alone VLSI designs to provide full functionality for an application.* This definition of SoC clearly emphasizes predesigned models of complex functions known as *cores* (terms such as intellectual property block, virtual components, and macros are also used) that serve a variety of applications. In SoC, an ASIC vendor may use a library of cores designed in-house as well as some cores from fabless/chipless design houses also known as intellectual property (IP) companies. The scenario for SoC design today is primarily characterized by three forms [1]:

1. *ASIC vendor design:* This refers to the design in which all the components in the chip are designed as well as fabricated by an ASIC vendor.

2. *Integrated design:* This refers to a design by an ASIC vendor in which all components are not designed by that vendor. It implies the use of one or multiple cores obtained from some other source such as a core/IP vendor or a foundry. The fabrication of these designs is done by either the ASIC vendor or a foundry company.

3. *Desktop design:* This refers to the design by a fabless company that uses cores which for the most part have been obtained from other

sources such as IP companies, EDA companies, design services companies, or a foundry. In the majority of cases, an independent foundry company fabricates these designs.

Because of the increasing integration of cores and the use of embedded software in SoC, the design complexity of SoC has increased dramatically and is expected to increase continuously at a very fast rate. Conceptually this trend is shown in Figure 1.1.

Every three years, silicon complexity quadruples following Moore's law. This complexity accounts for the increasing size of cores and the shrinking geometry that makes it necessary to include more and more parameters in the design criterion. For example, a few years ago it was sufficient to consider functionality, delay, power, and testability. Today, it is becoming increasingly important to also consider signal integrity, electromigration, packaging effects, electomagnetic coupling, and RF analysis.

In addition to the increasing silicon IP complexity, the embedded software content has increased at a rate much higher than that of Moore's law. Hence, on the same scale, overall system complexity has a much steeper slope than that of silicon complexity.

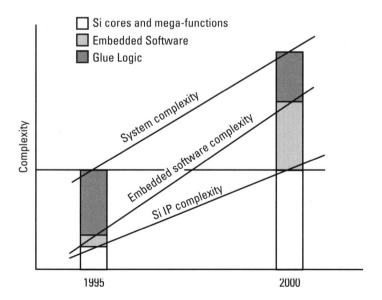

Figure 1.1 Trend toward increasing design complexity due to integration.

1.1 Architecture of the Present-Day SoC

In all SoC designs, predesigned cores are the essential components. A system chip may contain combinations of cores for on-chip functions such as microprocessors, large memory arrays, audio and video controllers, modems, Internet tuner, 2D and 3D graphics controllers, DSP functions, and so on. These cores are generally available in either synthesizable high-level description language (HDL) form such as in Verilog/VHDL, or optimized transistor-level layout such as GDSII. The flexibility in the use of cores also depends on the form in which they are available. Subsequently, soft, firm, and hard cores are defined as follows [1–3]:

- *Soft cores:* These are reusable blocks in the form of a synthesizable RTL description or a netlist of generic library elements. This implies that the user of soft core (macro) is responsible for the actual implementation and layout.

- *Firm cores:* These are reusable blocks that have been structurally and topologically optimized for performance and area through floor planning and placement, perhaps using a range of process technologies. These exist as synthesized code or as a netlist of generic library elements.

- *Hard cores:* These are reusable blocks that have been optimized for performance, power, and size, and mapped to a specific process technology. These exist as a fully placed and routed netlist and as a fixed layout such as in GDSII format.

The trade-off among hard, firm, and soft cores is in terms of parameters such as reusability, flexibility, portability, optimized performance, cost, and time-to-market. Qualitatively, this trade-off is shown in Figure 1.2.

The examples of core-based SoC include today's high-end microprocessors, media processors, GPS controllers, single-chip cellular phones, GSM phones, smart pager ASICs, and even PC-on-a-chip. Note that some people do not consider microprocessors within the definition of SoC; however, the architecture and design complexity of microprocessors such as the Alpha 21264, PowerPC, and Pentium III is no less than that of SoC by any measurement.

To understand the general architecture of SoC, Figure 1.3 shows an example of high-end microprocessors, and Figure 1.4 illustrates two SoC designs. Both figures show the nature of components used in today's SoC.

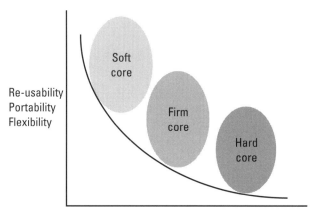

Re-usability
Portability
Flexibility

Higher predictability, performance, short SoC time-to-market
Higher cost and effort by the IP vendor

Figure 1.2 Trade-offs among soft, firm, and hard cores.

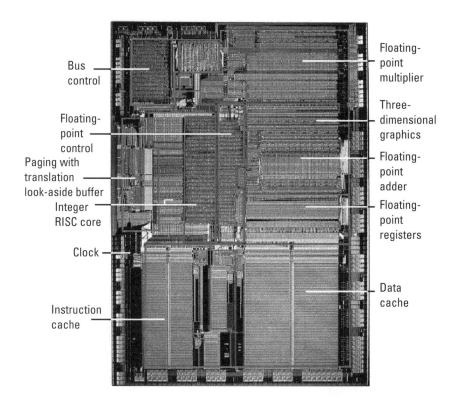

Figure 1.3 Intel's i860 microprocessor. (From [4], © IEEE 1989. Reproduced with permission.)

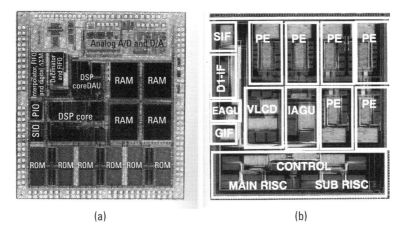

(a)   (b)

Figure 1.4 Examples of today's SoC: (a) Codec sign processor. (From [5], © IEEE 1996. Reprinted with permission.) (b) MPEG2 video coding/decoding. (From [6], © IEEE 1997. Reproduced with permission.)

Based on these examples, a generalized structure of SoC can be shown as given in Figure 1.5.

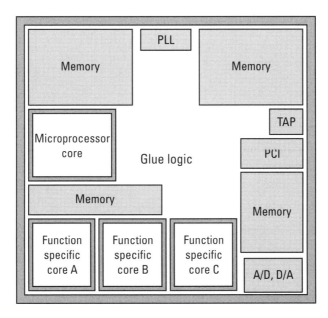

Figure 1.5 General architecture of today's embedded core-based system-on-a-chip.

Figures 1.3 to 1.5 illustrate examples of common components in today's SoC: multiple SRAM/DRAM, CAM, ROM, and flash memory blocks; on-chip microprocessor/microcontroller; PLL; sigma/delta and ADC/DAC functional blocks; function-specific cores such as DSP; 2D/3D graphics; and interface cores such as PCI, USB, and UART.

1.2   Design Issues of SoC

Due to the use of various hard, firm, and soft cores from multiple vendors, the SoC design may contain a very high level of integration complexity, interfacing and synchronization issues, data management issues, design verification, and test, architectural, and system-level issues. Further, the use of a wide variety of logic, memory, and analog/mixed-signal cores from different vendors can cause a wide range of problems in the design of SoC. In a recent survey by VLSI Research Inc., the following design issues were identified [7]:

Portability Methodology

- Non-netlisted cores;
- Layout-dependent step sizes;
- Aspect ratio misfits;
- Hand-crafted layout.

Timing Issues

- Clock redistribution;
- Hard core width and spacing disparities;
- Antenna rules disparities;
- RC parasitics due to chip layers;
- Timing reverification;
- Circuit timing.

Processing and Starting Material Difficulties

- Non-industry-standard process characteristics;
- *N*-well substrate connections;

- Substrate starting materials;
- Differences in layers between porting and target process.

Other Difficulties

- Mixed-signal designs are not portable;
- Accuracy aberrations in analog;
- Power consumption.

To address such a wide range of difficulties, a number of consortiums have developed (or are developing) guidelines for the design of cores and how to use them in SoC. Some notable efforts are:

- Pinnacles Component Information Standards (PCIS) by Reusable Application-Specific Intellectual Property Developers (RAPID) [8, 9];
- Electronic Component Information Exchange (ECIX) program by Silicon Integration Initiative (Si2) [10, 11]; and
- Embedded core design and test specifications by Virtual Socket Interface (VSI) Alliance [12–16].

The VSI Alliance has also developed an architecture document and specifications for an on-chip bus [12, 13]. The objectives of the architecture and on-chip bus (OCB) specifications are to accelerate the mix-and-match capabilities of cores. That is, in an SoC design with almost any on-chip bus, almost any virtual component interface (VCI) compliant core can be integrated. The conceptual view of a VSI OCB-based SoC design is illustrated in Figure 1.6 [13].

Conceptually, Figure 1.6 is similar to 1980s system design with a fixed interface such as an RS232, USB, or PCI bus. From a system design point of view, the components that support a common interface can be plugged into the system without significant problems using a fixed data transfer protocol.

Many companies have proposed proprietary bus-based architectures to facilitate core-based SoC design. Examples are IBM core-connect, Motorola IP-bus under M-Core methodology, ARM's advanced microcontroller bus architecture (AMBA), and advanced high-performance bus (AHB). The reason for this emphasis on OCB is that it permits extreme flexibility in core

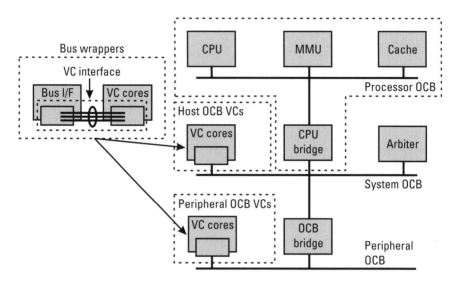

Figure 1.6 VSI hierarchical bus architecture for SoC design. (From [13], © VSIA 1998. Reproduced with permission.)

connectivity to OCBs by utilizing a fixed common interface across all cores. This architecture allows data and instruction flow from core-to-core and core-to-peripherals over on-chip buses. This is very similar to chip-to-chip communication in computers in the 1980s.

In terms of task responsibilities in SoC design, VSI defines its specifications as bridges between core provider and core integrator. An overview of this philosophy is illustrated in Figure 1.7 [3].

Most of the ASIC and EDA companies define flowcharts for design creation and standardize in-house design methodology based on that, from core design sign-off to SoC design sign-off. For example, IBM's Blue Book methodology and LSI Logic's Green Book methodologies are widely known. The web sites of most ASIC companies contain an overview of reuse/core-based design methodology and the specification of cores in their portfolio.

Traditionally, the front-end design of ICs begins with system definition in behavioral or algorithmic form and ends with floor planning, while the back-end design is defined from placement/routing through layout release (tape-out). Thus, the front-end design engineers do not know much about the back-end design process and vice versa. For effective SoC design, vertically integrated design engineers are necessary who have full responsibility for a block from system design specifications to physical design prior to chip-level integration. Such vertical integration is necessary for functional

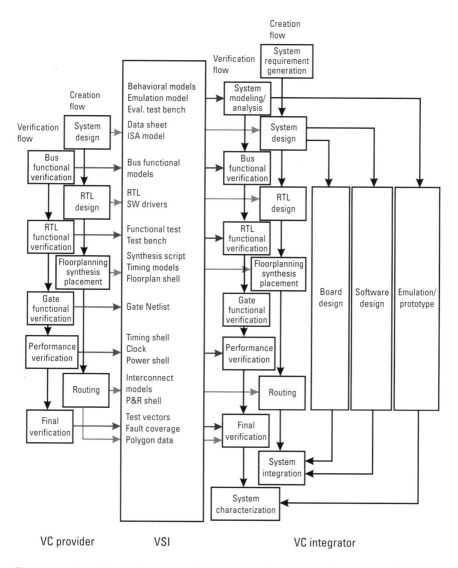

Figure 1.7 Virtual Socket Interface Alliance design flow for SoC. (From [3], © VSIA 1998. Reproduced with permission.)

verification of complex blocks with postlayout timing. This avoids last-minute surprises related to block aspect ratio, timing, routing, or even architectural and area/performance trade-offs.

In the present environment, almost all engineers use well-established RTL synthesis flow. In the general EDA synthesis flow, the designers translate the RTL description of the design to the gate level, perform various simulations at gate level to optimize the desired constraints, and then use EDA place and route flow. A major challenge these engineers face while doing SoC design is the description of functionality at the behavioral level in more abstract terms than the RT-level Verilog/VHDL description.

In a vertically integrated environment, design engineers are responsible for a wide range of tasks—from behavioral specs for RTL and mixed-signal simulation to floor planning and layout. An example of the task responsibilities of Motorola's Media Division engineers is shown in Figure 1.8 [17]. The necessary CAD tools used by this team for specific tasks are also shown in Figure 1.8.

In such a vertically-integrated environment, a large number of CAD tools are required and it is expected that most of the engineers have some knowledge of all the tools used by the team. To illustrate the complexity of the EDA environment used by SoC design groups, the list of tools supported by IBM under its Blue Logic Methodology is as follows [18]:

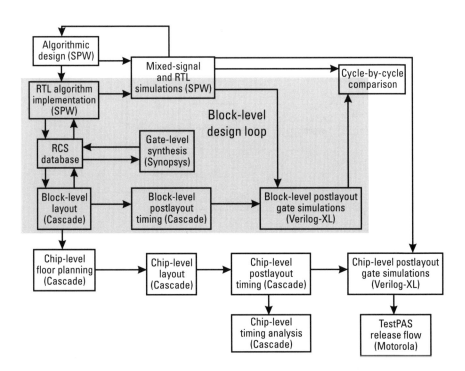

Figure 1.8 Task responsibilities of an engineer in a vertical design environment. (From [17], © IEEE 1997. Reproduced with permission.)

Design Flow

- *Schematic entry:* Cadence Composer, IBM Wizard.

- *Behavioral simulation:* Avanti Polaris and Polaris-CBS; Cadence Verilog-XL, Leapfrog, NC Verilog; Chronologic VCS; IBM TexSim; Mentor Graphics ModelSim; QuickTurn SpeedSim; Synopsys VSS.

- *Power simulation:* Sente Watt Watcher Architect; Synopsys Design-Power.

Technology Optimization

- *Logic synthesis:* Ambit BuildGates; IBM BooleDozer; Synopsys Design Compiler; DesignWare.

- *Power optimization:* Synopsys Power Compiler.

- *Front-end floor planning:* Arcadia Mustang; Cadence HLD Logic Design Planner; IBM ChipBench/HDP; Synopsys Floorplan Manager.

- *Clock planning:* IBM ClockPro.

- *Test synthesis:* IBM BooleDozer-Lite and DFTS; Logic Vision icBIST; Synopsys Test Compiler.

- *Clock synthesis netlist processing:* IBM BooleDozer-Lite and ClockPro.

Design Verification

- *Static timing analysis:* IBM EinsTimer; Synopsys DesignTime; Synopsys PrimeTime.

- *Test structure verification:* IBM TestBench, TSV and MSV.

- *Formal verification:* Chrysalis Design VERIFYer; IBM BoolesEye; Synopsys Formality.

- *Gate-level simulation:* Avanti Polaris and Polaris-CBS; Cadence Verilog-XL; Leapfrog; NC Verilog; Chronologic VCS; IBM TexSim; IKOS; Voyager-CS; Mentor Graphics ModelSim; QuickSim II; QuickTurn SpeedSim; Synopsys VSS.

- *Gate-level power estimation:* IBM PowerCalc; Synopsys Design-Power.
- *Prelayout technology checks:* IBM CMOS Checks.

Layout

- *Place and route:* IBM ASIC Design Center.
- *Technology checks:* IBM ASIC Design Center.
- *Automatic test pattern generation:* IBM ASIC Design Center.

Note that although the responsibilities shown in Figure 1.8 as well as knowledge of a large number of tools is required for high productivity of the SoC design team, this cross-pollination also enhances the engineers' knowledge and experience, overcomes communication barriers, and increases their value to the organization.

1.3 Hardware–Software Codesign

System design is the process of implementing a desired functionality using a set of physical or software components. The word *system* refers to any functional device implemented in hardware, software, or combinations of the two. When it is a combination of hardware and software, we normally call it *hardware–software codesign.* The SoC design process is primarily a hardware–software codesign in which design productivity is achieved by design reuse.

System design begins with specifying the required functionality. The most common way to achieve the precision in specification is to consider the system as a collection of simpler subsystems and methods for composing these subsystems (objects) to create the required functionality. Such a method is termed a *model* in the hardware–software codesign process. A model is formal; it is unambiguous and complete so that it can describe the entire system. Thus, a model is a formal description of a system consisting of objects and composition rules. Typically a model is used to decompose a system into multiple objects and then generate a specification by describing these objects in a selected language.

The next step in system design is to transform the system functionality into an architecture, which defines the system implementation by specifying the number and types of components and connections between them. The design process or methodology is the set of design tasks that transform an abstract specification model into an architectural model. Since we can have several possible models for a given system, selection of a model is based on system simulations and prior experience.

1.3.1 Codesign Flow

The overall process of system design (codesign) begins with identifying the system requirements. They are the required functions, performance, power, cost, reliability, and development time for the system. These requirements form the preliminary specifications often produced by the development teams and marketing professionals.

Table 1.1 provides a summary of some specification languages that can be used for system-level specifications and component functionality with respect to the different requirements of system designs. As the table shows, any one language is not adequate in all aspects of system specifications. VHDL, SDL, and JAVA seem to be the best choices. A number of publications describe these specification languages in substantial detail, and textbooks such as [19, 20] provide good overviews.

In terms of design steps, Figure 1.9 shows a generic codesign methodology flow at high level. Similar flows have been described in textbooks on codesign [19–22]. For a specific design, some of these steps may not be used or the flow may be somewhat modified. However, Figure 1.9 shows that simulation models are created at each step, analyzed and validated.

Table 1.1
Summary of System Specification Languages

Language	Concurrency	Communication	Timing	Interface	Note
VHDL	OK	Inadequate	Excellent	Text	IEEE standard
SDL	OK	OK	Inadequate	Text/graphics	ITU standard
Java	Excellent	Excellent	Inadequate	—	—
C, C++	N/A	N/A	N/A	Text	—
SpecChart	Excellent	OK	Excellent	—	—
StateChart	Excellent	Inadequate	OK	Graphics	—
PetriNet	Excellent	Inadequate	Excellent	Graphics	—
Esterel	Inadequate	Inadequate	Excellent	Text	—

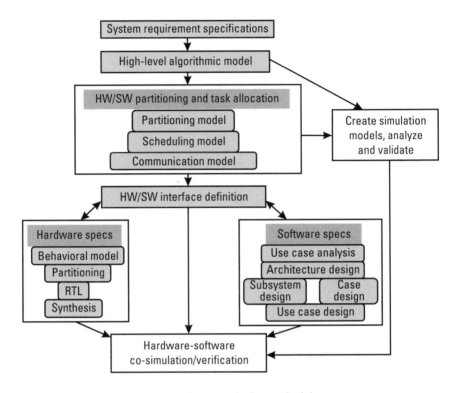

Figure 1.9 A general hardware–software codesign methodology.

Some form of validation and analysis is necessary at every step in order to reduce the risk of errors. The design steps include partitioning, scheduling, and communication synthesis, which forms the synthesis flow of the methodology. After these steps, a high-level algorithm and a simulation model for the overall system are created using C or C++. Some EDA tools such as COSSAP can be helpful in this process. With high-level algorithmic models, executable specs are obtained that are required by cosimulation. Because these specs are developed during the initial design phase, they require continuous refinement as the design progresses.

As the high-level model begins to finalize, the system architect decides on the software and hardware partitions to determine what functions should be done by the hardware and what should be achieved by the software applications.

Partitioning the software and hardware subsystems is currently a manual process that requires experience and a cost/performance trade-off. Tools such as Forsight are helpful in this task. The final step in partitioning is to define the interface and protocols between hardware and software followed by the detailed specs on individual partitions of both software and hardware.

Once the hardware and software partitions have been determined, a behavioral model of the hardware is created together with a working prototype of the software. The cosimulation of hardware and software allows these components to be refined and to develop an executable model with fully functional specs. These refinements continue throughout the design phase. Some of the major hardware design considerations in this process are clock tree, clock domains, layout, floor planning, buses, verification, synthesis, and interoperability issues. In addition, the entire project should have consistent rules and guidelines clearly defined and documented, with additional structures to facilitate silicon debugging and manufacturing tests.

Given a set of behaviors (tasks) and a set of performance constraints, scheduling is done to determine the order in which a behavior should run on a processing element (such as a CPU). In this scheduling the main considerations are (1) the partial order imposed by the dependencies in the functionality; (2) minimization of synchronization overhead between the processing elements; and (3) reduction of context switching overhead within the processing elements.

Depending on how much information about the partial order of behaviors is available at compile time, different scheduling strategies can be used. If any scheduling order of the behaviors is not known, then a run-time software scheduler can be used. In this case, the system model after the scheduling stage is not much different from the model after the partitioning stage, except that a new run-time software application is added for scheduling functionality.

On the other extreme, if the partial order is completely known at compile time, then a static scheduling scheme can be used. This eliminates context switching overhead of the behaviors, but it may suffer from interprocessing element synchronization, especially in the case of inaccurate performance estimation.

Up to the communication synthesis stage, communication and synchronization between concurrent behaviors are accomplished through shared variables. The task of the communication synthesis stage is to resolve the shared variable accesses into an appropriate interprocessing element communication at SoC implementation level. If the shared variable is a memory, the synthesizer will determine the location of such variables and change all accesses to this shared variable in the model into statements that read or write to the corresponding addresses. If the variable is in the local memory of one processing element, all accesses to this shared variable in the models of other processing elements have to be changed into function calls to message passing primitives such as send and receive.

The results of the codesign synthesis flow are fed to the back-end of the codesign process as shown in the lower part of Figure 1.9. If the hardware behavior is assigned to a standard processor, it will be fed into the compiler of this processor. This compiler should translate the design description into machine code for the target processor. If it is to be mapped into an ASIC, a high-level synthesis tool can synthesize it. The high-level synthesizer translates the behavioral design model into a netlist of RTL library components.

We can define interfaces as a special type of ASIC that links the processing elements associated (via its native bus) with other components of the system (via the system bus). Such an interface implements the behavior of a communication channel. For example, such an interface translates a read cycle on a processor bus to a read cycle on the system bus.

The communication tasks between different processing elements are implemented jointly by the driver routines and interrupt service routines implemented in software and interface circuitry implemented in hardware. While partitioning the communication task into hardware and software, the model generation for those two parts is the job of *communication synthesis*. The task of generating an RTL design from the interface model is the job of interface synthesis. The synthesized interface must synchronize the hardware protocols of the communicating components.

In summary, a codesign provides methodology for specification and design of systems that include hardware and software components. Hardware–software codesign is a very active research area. At the present time a set of tools is required because most of the commercial codesign tools are primarily cosimulation engines that do not provide system-level timing, simulation, and verification. Due to this lack of functionality in commercial tools, codesign presents a major challenge as identified in various case studies [23, 24]. In the future, we can expect to see the commercial application of specification languages, architectural exploration tools, algorithms for partitioning, scheduling in various synthesis stages in the flow, and back-end tools for custom hardware and software synthesis.

1.3.2 Codesign Tools

In recent years, a number of research groups have developed tools for codesign. Some of these tools are listed here:

- *Single processor architecture:* Cosyma [25, 26], Lycos [27], Mickey [28], Tosca [29], Vulcan [30];

- *Multiprocessor architecture:* Chinook [31], Cool [20, 32], Cosmos [33], CoWare [34], Polis [35], SpecSyn [36].

In addition to these tools, researchers have also developed system-modeling tools such as Ptolemy [37] and processor synthesis tools such as Castle [38]. Descriptions of these tools is beyond the scope of this book. However, to serve the purpose of an example, a brief overview of the Cosyma system is given.

Cosyma (co-synthesis for embedded microarchitecture) is an experimental system for design space exploration for hardware–software codesign (see Figure 1.10). It was developed in academic settings through multi-university cooperation. It shows where and how the automation of the codesign process can be accomplished. The target architecture of Cosyma consists of a standard RISC processor, RAM, and an automatically generated application-specific coprocessor. For ASIC development using these com- ponents, the peripheral units are required to be put in by the ASIC designer. The host processor and coprocessor communicate via shared memory [25, 26].

The system specs given to Cosyma consist of several communication processes written in a language derived from C (named Cx) in order to allow parallel processes. Process communication uses predefined Cx functions that access abstract channels, which are later mapped to physical channels or removed during optimization. Peripheral devices must be modeled in Cx for simulations. Cx is also used for stimulus generation. Both stimulus and peripheral models are removed for scheduling and partitioning. Another input is a list of constraints and a user directives file that contains time constraints referring to labels in Cx processes as well as channel mapping directives, partitioning directives, and component selections.

The input description is translated into an extended syntax graph after some analysis of local and global data flow of Cx processes. Then Cx processes are simulated on an RTL model of the target processor to obtain profiling and software timing information. This simulation step can be replaced by a symbolic analysis approach. Software timing data for each potential target processor is derived with simulation or symbolic analysis.

Multiple process systems then go through process scheduling steps to serialize the tasks. Cosyma considers data rates among processes for this pur-pose and uses partitioning and scheduling algorithms. The next step is to

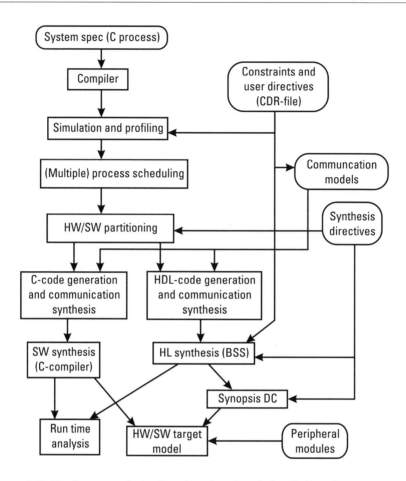

Figure 1.10   The Cosyma codesign flow, based on descriptions in [25, 26].

partition the processes (tasks) to be implemented in hardware or software. The inputs to this step are the extended syntax graph with profiling/control flow analysis data, CDR file, and synthesis directives. These synthesis directives include number and data of the functional units provided for coprocessor implementation. Also, they are needed to estimate the performance of the chosen/potential hardware configuration with the help of the user's interaction.

Partitioning is done at the basic block level in a Cx process. Partitioning requires communication analysis and communication synthesis. Some other codesign tools/flows require that the user provide explicit communication channel information and then partition at the level of the Cx processes.

Cosyma inserts communication channels when it translates the extended syntax graph representation back to C code for software synthesis and to a HDL for high-level hardware synthesis.

For high-level synthesis, the Braunschweig Synthesis System (BSS) is used. BSS creates a diagram showing the scheduling steps, function units, and memory utilization, which allow the designer to identify bottlenecks. The Synopsys Design compiler creates the final netlist. The standard C compiler helps in software synthesis from the Cx process partitions. The run-time analysis step includes hardware–software cosimulation using the RTL hardware code.

1.4 Core Libraries, EDA Tools, and Web Pointers

Before concluding this chapter, it is worth mentioning that an enormous amount of information on SoC is available on the web. By no means can this chapter or book capture that information. This section serves merely as a guide to core libraries and EDA tools and provides some web pointers to company web sites for readers interested in further information.

1.4.1 Core Libraries

A number of companies have developed core libraries. The cores in such libraries are generally optimized and prequalified on specific manufacturing technologies. These libraries contain cores that implement a wide range of functions from microprocessors/microcontrollers, DSP, high-speed communication controllers, memories, bus functions and controllers, and analog/mixed-signal circuits such as PLL, DAC/ADC, and so on. As an example, a summary of LSI Logic's core library is given as follow [39]:

- TinyRISC 16/32-bit embedded TR4101 CPU
- TinyRISC 16/32-bit embedded TR4101 CPU embedded in easy macro (EZ4102);
- MiniRISC 32-bit superscaler embedded CW4003 CPU;
- MiniRISC 32-bit superscaler embedded CW4011 CPU;
- MiniRISC 64-bit embedded CW40xx CPU;
- Oak DSPCore CPU 16-bit fixed-point CWDSP1640;
- Oak DSPCore CPU 16-bit fixed-point CWDSP1650;
- GigaBlaze transceiver;

- Merlin fiber channel protocol controller;
- Viterbi decoder;
- Reed-Solomon decoder;
- Ethernet-10 controller (include 8-wire TP-PMD), 10 Mbps;
- MENDEC-10 Ethernet Manchester encoder-decoder, 10 Mbps;
- Ethernet-I 10 MAC, 10/100 Mbps;
- SONET/SDH interface (SSI), I 55/5 I Mbps;
- ARM7 thumb processor;
- T1 framer;
- HDLC;
- Ethernet-I 10 series, 10/100 Mbps;
- Ethernet-I 10 100 base-x, 10/100 Mbps;
- PHY-I 10, Ethernet auto negotiation 10/1000 Mbps;
- USB function core;
- PCI-66 FlexCore;
- 1-bit slicer ADC 10 MSPS;
- 4-bit low power flash DC 10 MSPS;
- 6-bit flash ADC 60 MSPS;
- 6-bit flash ADC 90 MSPS;
- 8-bit flash ADC 40 MSPS;
- 10-bit successive approximation ADC 350 KSPS;
- Triple 10-bit RGB video DAC;
- 10-bit low-power DAC 10 MSPS;
- 10-bit low-power multiple output DAC;
- Sample-and-hold output stage for 10-bit low-power multiple output DAC;
- Programmable frequency synthesizer 300 MHz;
- SONET/ATM 155 MSPS PMD transceiver;
- 155 and 207 MBPS high-speed backplane transceiver;
- Ethernet 10Base-T/A UI 4/6 pin, 5V;
- Ethernet 100Base-x clock generation/data recovery functions, 3V.

1.4.2 EDA Tools and Vendors

The EDA vendors provide a large number of design automation tools that are useful in SoC design. This list is not complete and does not imply any endorsement. The web site of *Integrated System Design* magazine (http://www.isdmag.com/design.shtml) contains a number of articles with extensive surveys on tools. In most cases, the exact description of a tool can be obtained from the company web site.

Codesign

- Comet from Vast Systems;
- CVE from Mentor Graphics;
- Foresight from Nutherma Systems;
- Eagle from Synopsys;
- CosiMate (system level verification) and ArchiMate (architecture generation) from Arexsys.

Design Entry

- Discovery (interactive layout), Nova-ExploreRTL (Verilog, VHDL) from Avanti;
- Cietro (system-level design in graphics, text, C, HDL, Matlab, FSM) and Composer from Cadance;
- SaberSketch (mixed-signal circuits in MAST, VHDL-AMS and C) from Analogy;
- Quickbench from Chronology;
- RADware Software from Infinite Technology;
- Debussy from Novas Software;
- QuickWorks from QuickLogic;
- EASE and EALE from Translogic;
- Origin (data management) and VBDC from VeriBest;
- ViewDraw from Viewlogic;
- Wizard from IBM.

Logic Simulation

- VerilogXL (Verilog), LeapFrog(VHDL), Cobra (Verilog), Affirma NC Verilog, Affirma NC VHDL, Affirma Spectre (analog, mixed-signal), Affirma RF simulation and Affirma Verilog-A (behavioral Verilog) from Cadence Design Systems;

- Quickbench Verification Suite (Verilog, VHDL) from Chronology;

- VSS(VHDL), VCS (Verilog), TimeMill (transistor-level timing simulator), Vantage-UltraSpec (VHDL) and Cyclone (VHDL), CoverMeter (Verilog) from Synopsys;

- V-System (VHDL/Verilog) from Model Technology;

- PureSpeed (Verilog) from FrontLine Design Automation (now Avanti), Polaris and Polaris-CBS from Avanti;

- TexSim from IBM;

- ModelSim (Verilog, VHDL), Seamless CVE (cosimulation) from Mentor Graphics;

- SpeedSim (VHDL/Verilog) from Quickturn design systems Inc.;

- FinSim-ECST from Fintronic USA Inc.;

- PeakVHDL from Accolade Design Automation;

- VeriBest VHDL, VeriBest Verilog, VBASE (analog, mixed A/D) from VeriBest;

- Fusion Speedwave (VHDL), Fusion VCS (Verilog), Fusion View-Sim (digital gate-level) from Viewlogic.

Formal Verification Tools

- Formality from Synopsys;

- Affirma Equivalence Checker from Cadence;

- DesignVerifier and Design Insight from Chrysalis;

- CheckOff/Lambda from Abstract Inc.;

- LEQ Logic Equivalency and Property Verifier from Formalized Design Inc.;

- Tuxedo from Verplex Systems;

- Structureprover II from Verysys Design Automation;

- VFormal from Compass Design Automation (Avanti Corporation);

- FormalCheck from Bell Labs Design Automation.;

- BooleEye and Rulebase from IBM.

Logic Synthesis Tools

- Design Compiler (ASIC), Floorplan Manager, RTL Analyzer and FPGA-Express (FPGA) from Synopsys;
- BuildGates from Ambit Design Systems (Cadence);
- Galileo (FPGA) from Exemplar (Mentor Graphics);
- Symplify (FPGA), HDL Analyst and Certify (ASIC prototyping in multiple FPGAs) from Symplicity Inc.;
- RADware Software from Infinite Technology;
- Concorde (front end RTL synthesis), Cheetah (Verilog), Jaguar (VHDL) and NOM (development system support) from Interra;
- BooleDozer (netlist), ClockPro (clock synthesis) from IBM.

Static Timing Analysis Tools

- PrimeTime (static), DesignTime, Motive, PathMill (static mixed level), CoreMill (Static transistor level), TimeMill (dynamic transistor level), DelayMill (static/dynamic mixed level) from Synopsys;
- Saturn, Star RC (RC extraction), Star DC and Star Power (power rail analysis) from Avanti;
- TimingDesigner (static/dynamic) from Chronology;
- Path Analyzer (static) from QuickLogic;
- Pearl from Cadence Design Systems;
- Velocity (static) from Mentor Graphics;
- BLAST (static) from Viewlogic;
- EinsTimer from IBM.

Physical Design Parasitic Extraction Tools

- HyperExtract from Cadence Design Systems;
- Star-Extract from Avanti Corporation;
- Arcadia from Synopsys;
- Fire&Ice from Simplex Solutions.

Physical Design

- HDL Logic Design Planner, Physical design planner, SiliconEnsemble, GateEnsemble, Assura Vampire, Assura Dracula, Virtuoso and Craftsman from Cadence Design Systems;
- Hercules, Discovery, Planet-PL, Planet-RTL and Apollo from Avanti Corporation;
- Floorplan Manager, Cedar, Arcadia, RailMill from Synopsys;
- Blast Fusion from Magma Design Automation;
- MCM Designer, Calibre, IS Floorplanner, IS Synthesizer and IC station from Mentor Graphics;
- Dolphin from Monterey Design Systems;
- Everest System from Everest Design Automation;
- Epoch from Duet Technologies;
- Cellsnake and Gatesnake from Snaketech Inc.;
- Tempest-Block and Tempest-Cell from Sycon Design Inc.;
- L-Edit Pro and Tanner Tools Pro from Tanner EDA;
- Columbus Interconnect Modeler, Columbus Inductance Modeler, Cartier Clock Tree Analyzer from Frequency Technology;
- RADware Software from Infinite Technology;
- CircuitScope from Moscape;
- Dream/Hurricane, Companion and Xtreme from Sagantec North America;
- ChipBench/HDP (floorplan), ClockPro (clock plan) from IBM;
- Grandmaster and Forecast Pro from Gambit Design Systems;
- Gards and SonIC from Silicon Valley Research Inc.

Power Analysis Tools

- DesignPower, PowerMill and PowerCompiler from Synopsys;
- Mars-Rail and Mars Xtalk from Avanti;
- CoolIt from InterHDL;
- WattWatcher from Sente Inc.;
- PowerCalc from IBM.

ASIC Emulation Tools

- Avatar and VirtuaLogic, VLE-2M and VLE-5M from IKOS systems Inc.;

- SimExpress from Mentor Graphics;

- Mercury Design Verification and CoBALT from Quickturn Systems Inc.;

- System Explorer MP3C and MP4 from Aptix.

Test and Testability Tools

- Asset Test Development Station, Asset Manufacturing Station and Asset Repair Station from Asset Intertech Inc.;

- Faultmaxx/Testmaxx, Test Design Expert, Test Development Series and BISTmaxx from Fluence Technology;

- LogicBIST, MemBIST, Socketbuilder, PLLBIST, JTAG-XLI from LogicVision;

- Fastscan, DFTadvisor, BSDarchitect, DFTinsight, Flextest, MBIS-Tarchitect and LBISTarchitect from Mentor Graphics;

- Teramax ATPG, DC Expert Plus and TestGen from Synopsys;

- TurboBIST-SRAM, TurboBSD, Turbocheck-RTL, Turbocheck-Gate, TurboFCE, Turboscan and Turbofault from Syntest Technologies;

- FS-ATG test vector generation and FS-ATG Boundary Scan test generation from Flynn System Corporation;

- Intellect from ATG Technology;

- Eclipse scan diagnosis from Intellitech Corporation;

- Test Designer from Intusoft;

- Testbench from IBM;

- Verifault from Cadence;

- Hyperfault from Simucad;

- Testify from Analogy.

1.4.3   Web Pointers

Some useful URLs are listed next for readers seeking additional information:

Guides, News, and Summaries

- Processors and DSP guides, http://www.bdti.com/library.html;
- Design and Reuse Inc., http://www.us.design-reuse.com;
- Integrated System Design, http://www.isdmag.com/sitemap.html;
- EE Times, http://www.eet.com/ipwatch/.

Company Sites

- Advance Risc Machine (ARM), http://www.arm.com;
- Altera MegaCores,
  http://www.altera.com/html/products/megacore.html;
- DSP Group, http://www.dspg.com/prodtech/core/main.htm;
- Hitachi, http://semiconductor.hitachi.com/;
- IBM, http://www.chips.ibm.com/products/,
  http://www.chips.ibm.com/bluelogic/;
- LogicVision, http://www.lvision.com/products.htm;
- LSI Logic, http://www.lsil.com/products/unit5.html;
- Lucent Technology, http://www.lucent.com/micro/products.html;
- Mentor Graphics, http://www.mentorg.com/products/;
- Mentor Graphics Inventra, http://www.mentorg.com/inventra/;
- National Semiconductor, http://www.nsc.com/diagrams/;
- Oak Technology, http://www.oaktech.com/technol.htm;
- Palmchip, http://www.palmchip.com/products.htm;
- Philips, http://www-us2.semiconductors.philips.com/;
- Phoenix Technology, http://www.phoenix.com/products/;
- Synopsys,
  http://www.synopsys.com/products/designware/8051_ds.html;
  http://www.synopsys.com/products/products.html;
- Texas Instruments, http://www.ti.com/sc/docs/schome.htm;
- Virtual Chips synthesizable cores, http://www.vchips.com;

- Xilinx, http://www.xilinx.com/products/logicore/logicore.htm;
- Zilog, http://www.zilog.com/frames/fproduct.html.

Standards Organizations

- RAPID, http://www.rapid.org;
- VSI Alliance, http://www.vsi.org;
- Silicon Initiative, Inc. (Si2), http://www.si2.org.

References

[1] Rincon, A. M., C. Cherichetti, J. A. Monzel, D. R. Stauffer, and M. T. Trick, "Core design and system-on-a-chip integration," *IEEE Design and Test of Computers,* Oct.–Dec. 1997, pp. 26–35.

[2] Hunt, M., and J. A. Rowson, "Blocking in a system on a chip," *IEEE Spectrum,* Nov. 1996, pp. 35–41.

[3] VSI Alliance, "Overview document," 1998.

[4] Perry, T. S., "Intel's secret is out," *IEEE Spectrum,* 1989, pp. 22–28.

[5] Norsworthy, S. R., L. E. Bays, and J. Fisher, "Programmable CODEC signal processor," *Proc. IEEE Int. Solid State Circuits Conf.,* 1996, pp. 170–171.

[6] Iwata, E., et al., "A 2. 2GOPS video DSP with 2-RISC MIMD, 6-PE SIMD architecture for real-time MPEG2 video coding/decoding," *Proc. IEEE Int. Solid State Circuits Conf.,* 1997, pp. 258–259.

[7] Hutcheson, J., "Executive advisory: The market for systems-on-a-chip," June 15, 1998, and "The market for systems-on-a-chip testing," July 27, 1998, VLSI Research Inc.

[8] Reusable Application-Specific Intellectual Property Developers (RAPID) web site, http://www.rapid.org.

[9] Glover, R., "The implications of IP and design reuse for EDA," *EDA Today,* 1997.

[10] Si2, "The ECIX program overview," 1998.

[11] Cottrell, D. R., "ECIX: Electronic component information exchange," Si2, 1998.

[12] VSI Alliance Architecture document, version 1.0, 1997.

[13] VSI Alliance, "On-chip bus attributes," OCB 1 1.0, August 8, 1998.

[14] VSI Alliance system level design taxonomy and terminology, 1998.

[15] Analog/mixed-signal VSI extension, VSI Alliance Analog/Mixed-Signal Extension document, 1998.

[16] Structural netlist and hard VS physical data types," VSI Implementation/Verification DWG document, 1998.

[17] Eory, F. S., "A core-based system-to-silicon design methodology," *IEEE Design and Test of Computers,* Oct.–Dec. 1997, pp. 36–41.

[18] IBM Microelectronic web site, http://www.chips.ibm.com/bluelogic/.

[19] Jerraya, A., et al., "Languages for system-level specification and design," in *Hardware/Software Codesign: Principles and Practices,* Norwell, MA: Kluwer Academic Publishers, 1997, pp. 36–41.

[20] Niemann, R., *Hardware/Software Codesign for Data Flow Dominated Embedded Systems,* Norwell, MA: Kluwer Academic Publishers, 1998.

[21] van den Hurk, J., and J. Jess, *System Level Hardware/Software Codesign,* Norwell, MA: Kluwer Academic Publishers, 1998.

[22] Keating, M., and P. Bricaud, *Reuse Methodology Manual,* Norwell, MA: Kluwer Academic Publishers, 1998.

[23] Cassagnol, B., et al., "Codesigning a complex system-on-a-chip with behavioral models," *Integrated Systems Design,* Nov. 1998, pp. 19–26.

[24] Adida, C., et al., "Hardware–software codesign of an image processing unit," *Integrated Systems Design,* July 1999, pp. 37–44.

[25] Cosyma ftp site, ftp://ftp.ida.ing.tu-bs.de/pub/cosyma.

[26] Osterling, A., et al., "The Cosyma system," in *Hardware/Software Codesign: Principles and Practices,* pp. 263–282, Kluwer Academic Publishers, 1997.

[27] Madsen, J., et al., "LYCOS: The lyngby co-synthesis system," *Design Automation for Embedded Systems,* Vol. 2, No. 2, 1997, pp. 195–235.

[28] Mitra, R. S., et al., "Rapid prototyping of microprocessor based systems," *Proc. Int. Conf. on Computer-Aided Design,* 1993, pp. 600–603.

[29] Balboni, A., et al., "Co-synthesis and co-simulation of control dominated embedded systems," *Design Automation for Embedded Systems,* Vol. 1, No. 3, 1996. pp. 257–289.

[30] Gupta, R. K., and G. De Micheli, "A co-synthesis approach to embedded system design automation," *Design Automation for Embedded Systems,* Vol. 1, Nos. 1–2, 1996, pp. 69–120.

[31] Chao, P., R. B. Ortega, and G. Boriello, "Interface co-synthesis techniques for embedded systems," *Proc. Int. Conference on Computer-Aided Design,* 1995, pp. 280–287.

[32] Niemann, R., and P. Marwedel, "Synthesis of communicating controllers for concurrent hardware/software systems," *Proc. Design Automation and Test in Europe,* 1998.

[33]   Ismail, T. B., and A. A. Jerraya, "Synthesis steps and design models for codesign," *IEEE Computer,* 1995, pp. 44–52.

[34]   van Rompaey, K., et al., "CoWare—A design environment for heterogeneous hardware/software systems," *Proc. European Design Automation Conference,* 1996.

[35]   Chiodo, M., et al., "A case study in computer aided codesign of embedded controllers," *Design Automation for Embedded Systems,* Vol. 1, Nos. 1–2, 1996, pp. 51–67.

[36]   Gajski, D., F. Vahid, and S. Narayanan, "A system design methodology: executable specification refinement," *European Design and Test Conference,* 1994, pp. 458–463.

[37]   Kalavade, A., and E. A. Lee, "A hardware–software codesign methodology for DSP applications," *IEEE Design and Test,* 1993, pp. 16–28.

[38]   Wilberg, J., and R. Camposano, "VLIW processor codesign for video processing," *Design Automation for Embedded Systems,* Vol. 2, No. 1, 1997, pp. 79–119.

[39]   LSI Logic web site, http://www.lsil.com/products/unit5.html.

2

Design Methodology for Logic Cores

To maintain productivity levels when dealing with ever-increasing design complexity, design-for-reuse is an absolute necessity. In cores and SoC designs, design-for-reuse also helps keep the design time within reasonable bounds. Design-for-reuse requires good functional documentation, good coding practices, carefully designed verification environments, thorough test suites, and robust and versatile EDA tool scripts. Hard cores also require an effective porting mechanism across various technology libraries.

A core and its verification testbench targeted for a single HDL language and a single simulator are generally not portable across the technologies and design environments. A reusable core implies availability of verifiably different simulation models and test suites in several major HDLs, such as Verilog and VHDL. Reusable cores must have stand-alone verification testbenches that are complete and can be simulated independently.

Much of the difficulty surrounding the reuse of cores is also due to inadequate description of the core, poor or even nonexistent documentation. Particularly in the case of hard cores, a detailed description is required of the design environment in which the core was developed as well as a description of the simulation models. Because a core provider cannot develop simulation models for all imaginable uses, many times SoC designers are required to develop their own simulation models of the core. Without proper documentation, this is a daunting task with a high probability of incomplete or erroneous functionality.

2.1 SoC Design Flow

SoC designs require an unconventional design methodology because pure top-down or bottom-up design methodologies are not suitable for cores as well as SoC. The primary reason is that during the design phase of a core, all of its possible uses cannot be conceived. A pure top-down design methodology is suitable when the environment in which the core will be used is known a priori and that knowledge is used in developing the functional specifications. Because of the dependency on the core design, the SoC design methodology is a combination of bottom-up and top-down philosophies that look like an interlaced model based on hardware–software codevelopment while simultaneously considering physical design and performance. This design methodology is considerably different than the traditional ASIC design philosophy in which design tasks are done in sequential order.

Such design flow is described in a horizontal/vertical model as shown in Figure 2.1. Similar flows have been mentioned in the literature [1, 2]. In such a design flow, although the architectural design is based on hardware–software codevelopment, the VLSI design requires simultaneous analysis and optimization of area, performance, power, noise, test, technology constraints, interconnect, wire loading, electromigration, and packaging constraints. Because SoC may also contain embedded software, the design methodology also requires that the both hardware and software be developed concurrently to ensure correct functionality. Hardware–software codesign was briefly mentioned in Chapter 1 (Section 1.3.1) and illustrated in Figure 1.9.

The first part in this design process consists of recursive development and verification of a set of specifications until it is detailed enough to allow RTL implementation. This phase also requires that any exceptions, corner cases, limitations, and so on be documented and shared with everyone directly involved in the project. The specifications should be independent of the implementation method. There are two possible ways to develop specifications: formal specifications and simulatable specifications.

Formal specifications can be used to compare the implementation at various levels to determine the correctness from one abstraction level to another [3, 4], such as through the use of equivalence and property checking [5]. A few formal specification languages such as VSPEC [6] have been developed to help in specifying functional behavior, timing, power consumption, switching characteristics, area constraints, and other parameters. However, these languages are still in their infancy and robust commercial tools for formal specifications are not yet available. Today, simulatable specifications are

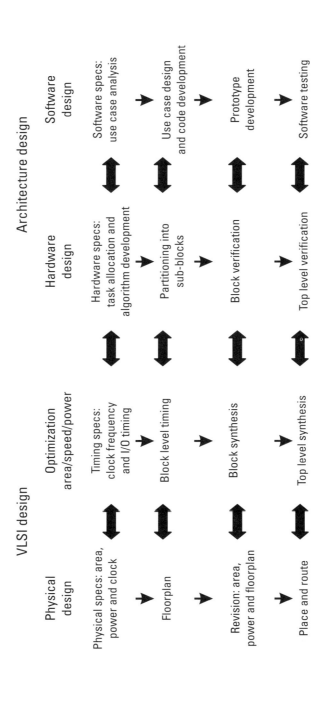

Figure 2.1 Interlaced horizontal/vertical codevelopment design methodology.

most widely used. Simulatable specifications describe the functional behavior of the design in an abstract form and do not provide a direct link from high-level specs to the RT level. Simulatable specifications are basically executable software models written in C, C++, or SDL, while the hardware is specified in Verilog or VHDL.

2.2 General Guidelines for Design Reuse

A number of precautions must be taken at various design steps to ensure design reusability. Some of these precautions are basic common sense while others are specific architectural or physical design guidelines.

2.2.1 Synchronous Design

Synchronous design style is extremely useful for core-based SoC design. In synchronous design, data changes based on clock edges only (and, hence, instructions and data) are easily manageable. Use of registers in random logic as well as registration at the inputs and outputs of every core as shown in Figure 2.2 is very useful in managing core-to-core interaction. Such registration essentially creates a wrapper around a core. Besides providing synchronization at the core boundary, it also has other benefits such as portability and application of manufacturing test. (Test aspects will be discussed in Chapter 6.)

Latch-based designs on the other hand are not easy to manage because the data capture is not based on a clock edge; instead, it requires a longer period of an active signal. It is thus useful to avoid latches in random logic and use them only in blocks such as FIFOs, memories, and stacks. In general, asynchronous loops and internal pulse generator circuits should be avoided in the core design. Similarly, multicycle paths and direct combinational paths from block inputs to outputs should be avoided. If there are any asynchronous clear and set signals, then their deactivation should be resynchronized. Furthermore, the memory boundaries at which read, write, and enable signals are applied should be synchronous and register-based.

2.2.2 Memory and Mixed-Signal Design

The majority of embedded memories in SoC are designed using memory compilers. This topic is discussed in detail in Chapter 3. While the memory design itself is technology dependent, some basic rules are very useful in SoC-level integration.

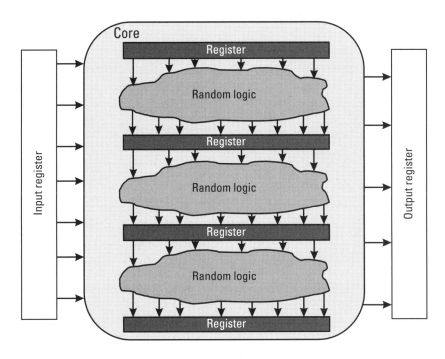

Figure 2.2 Use of registers for synchronization in core logic and its inputs and outputs.

In large memories, the parasitics at the boundary cell are substantially different than the parasitics of a cell in the middle of an array. To minimize this disparity, it is extremely useful to include rows and columns of dummy cells at the periphery of large memories as shown in Figure 2.3(a). To minimize the area overhead penalty because of these dummy cells, these rows and columns should be made part of the built-in self-repair (BISR) mechanism. BISR allows a bad memory cell to be replaced and also improves the manufacturing yield. A number of BISR schemes are available and many are discussed in Chapter 3.

While the large memories are generally placed along the side or corner of the chip, small memories are scattered all over the place. If not carefully planned, these small memories create a tremendous hurdle in chip-level routing. Hence, when implementing these small memories, it is extremely useful for the metal layers to be kept to one or two metals less than the technology allowable layers. Subsequently, these metals can be used to route chip-level wires over the memories.

In present-day SoC design, in general, more than 60% of the chip is memories; mixed-signal circuits make up hardly 5% of the chip area [7]. The

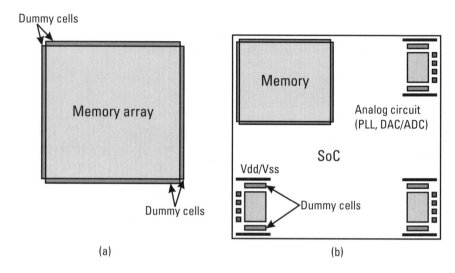

Figure 2.3 (a) Use of dummy cells with memory array. (b) Placement of memory and analog circuits at SoC level.

most commonly used analog/mixed-signal circuits used in SoC are PLLs, digital-to-analog converters (DACs), analog-to-digital converters (ADCs), and temperature sensors. These circuits provide specialized functionality such as on-chip clock generation, synchronization, RGB output for color video display, and communication with the outside world. Because these blocks are analog/mixed-signal circuits, they are extremely sensitive to noise and technology parameters. Thus, it is useful to place these circuits at the corners as shown in Figure 2.3(b). This also suggests that placing the I/Os of the analog circuit on only two sides is somewhat useful in simplifying their placement at the SoC level. Further, the use of guard-bands and dummy cells around these circuits (as shown in Figure 2.3) is useful to minimize noise sensitivity.

2.2.3 On-Chip Buses

On-chip buses play an extremely important role in SoC design. Bus-based designs are easy to manage primarily because on-chip buses provide a common interface by which various cores can be connected. Thus, the design of on-chip buses and the data transaction protocol must be considered prior to the core selection process. On-chip bus design after the selection and development of cores leads to conflicting data transfer mechanisms. Subsequently, it causes complications at SoC-level integration and results in additional hardware as well as lower performance.

Because core providers cannot envision all possible interfaces, parameterized interfaces should be used in the core design. For example, FIFO-based interfaces are reasonably flexible and versatile in their ability to handle varying data rates between cores and the system buses. A number of companies and organizations such as VSI Alliance are actively working to develop an acceptable on-chip bus and core interface standard/specifications that support multiple masters, separate identity for data and control signals, fully synchronous and multiple cycle transactions, bus request-and-grant protocol.

2.2.4 Clock Distribution

Clock distribution rules are one of the most important rules for cores as well as SoC designs. Any mismatch in clocking rules can impact the performance of an entire SoC design. It may even cause timing failures throughout the design. Therefore, establishing robust clock rules is necessary in SoC design. These rules should include clock domain analysis, style of clock tree, clock buffering, clock skew analysis, and external timing parameters such as setup/hold times, output pin timing waveforms, and so on. The majority of SoCs consist of multiple clock domains; it is always better to use the smallest number of clock domains. It is better to isolate each clock in an independent domain and use buffers at the clock boundary.

If two asynchronous clock domains interact, the interaction should be limited to a single, small submodule in the design hierarchy. The interface between the clock domains should avoid metastability and the synchronization method should be used at the clock boundaries. A simple resynchronization method consists of clock buffering and dual stage flip-flops or FIFOs at the clock boundary.

When cores contain local PLLs, a low-frequency chip-level synchronization clock should be distributed with on-chip buses. Each core's local PLL should lock to this chip-level synchronization clock and generate required frequency for the core.

Control on clock skew is an absolute necessity in SoC design. It avoids data mismatch as well as the use of data lock-up latches. A simple method to minimize clock skew is to *edge-synchronize* master and derived clocks. The general practice has been to use a balanced clock tree that distributes a single clock throughout the chip to minimize the clock skew. Examples of such trees are given in Figure 2.4. The basic principle is to use a balanced clock

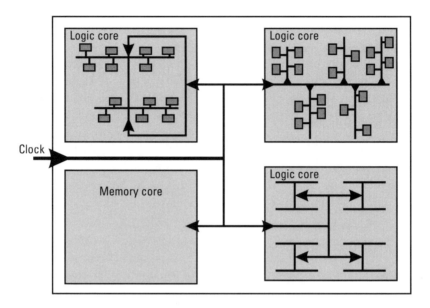

Figure 2.4 Clock distribution schemes for balanced load and minimized clock skew.

tree and clock buffer at the beginning of the clock tree so that any skew at the upper level can be adjusted by adjusting the buffer delay.

2.2.5 Clear/Set/Reset Signals

It is essential to document all reset schemes in detail for the entire design. The documentation should state whether resets are synchronous, asynchronous, or internal/external power-on-resets, how many resets are used, any software reset schemes used, whether any functional block has its locally generated resets, whether resets are synchronized with local clocks, and so on. Whenever possible, synchronous reset should be used because it avoids race conditions on reset. Static timing analysis becomes difficult with asynchronous resets, and the designer has to carefully evaluate the reset pulse width at every flip-flop to make sure it becomes inactive synchronously to clocks. Hence, whenever reset/clear is asynchronous, their deactivation should be resynchronized.

2.2.6 Physical Design

A number of physical design issues are extremely important from the reuse point of view. In the development of hard cores, physical design is a key item for the success of the core. Although soft and firm cores are not delivered in layout form, consideration of their physical design issues is still necessary.

2.2.6.1 Floor Plan

Floor planning should start early in the design cycle. It helps in estimating the size and in determining if area, timing, performance, and cost goals can be satisfied. The initial floor plan also helps in determining the functional interfaces among different cores as well as clock distribution at the chip level. When a SoC combines hard and soft cores, the fixed-aspect ratio of the hard core can impose placement and routing constraints on the rest of the design. Therefore, a low-effort SoC-level floor planning should be done in the early design process.

2.2.6.2 Synthesis

The overall synthesis process should also be planned early in the design phase and should include specific goals for area, timing, and power. Present-day synthesis tools do not handle very large design all at once; hence, hierarchically incremental synthesis should be done. For this, whole design should be partitioned into blocks small enough to be used by EDA tools [8, 9]. However, in this process each block should be floor-planned as a single unit to maintain the original wire load model.

Chip-level synthesis then consists of connecting various blocks and resizing the output drive buffers to meet the actual wire load and fan-out constraints. Hence, each block at this level should appear as two modules (in hierarchy), one enclosing the other, similar to a wrapper. The outer module contains output buffers and it can be incrementally compiled by the synthesis tool, whereas the inner module that contains functional logic of the core is not to be changed ("don't touch") by the tool. This type of synthesis wrapper ensures that the gate-level netlist satisfies area, speed, and power constraints.

2.2.6.3 Timing

Static timing analysis should be done before layout on floor-planned blocks. The final timing verification should be done on postlayout blocks. During timing analysis careful attention should be paid to black boxing, setup/hold time checks, false path elimination, glitch/hazard detection, loop removal, margin analysis, min/max analysis, multipath analysis, and clock skew analysis. This timing analysis should be repeated over the entire range of PVT (process, voltage, and temperature) specifications. Similar to the synthesis wrapper, a timing wrapper should be generated for each block. This timing wrapper provides a virtual timing representation of the gate-level netlist to other modules in the hierarchy.

2.2.6.4   Inputs/Outputs

The definition of core I/Os is extremely important for design reuse. The configuration of each core I/O, whether it is a clock input or a test I/O, should be clearly specified. These specifications should include type of I/Os (input/output/bidirect signal, clock, Vdd/Gnd, test-related I/O, as well as dummy I/Os), timing specifications for bidirect enable signals, limits on output loading (fan-out and wire load), range of signal slew rate for all inputs, and noise margin degradation with respect to capacitive load for outputs.

Placement of I/Os during the core design phase is also important because their placement impacts core placement in SoC. As a rule of thumb, all power/ground pins of the cores should be placed on one side so that when a core is placed on one side of the chip, these pins become chip-level power/ground pins. This rule is a little tricky for signal I/Os. However, placing signal I/Os at two sides of the core (compared to distributing along all four sides) is beneficial in the majority of cases.

2.2.6.5   Validation and Test

Design validation and test are critical for successful design reuse. However, this discussion is skipped here; validation is addressed in Chapter 4, while test methodologies and manufacturing test are discussed in Chapters 6 to 10.

2.2.7   Deliverable Models

The reuse of design is pretty much dependent on the quality of deliverable models. These models include a behavioral or instruction set architecture (ISA) model, a bus functional model for system-level verification, a fully functional model for timing and cycle-based logic simulation/emulation, and physical design models consisting of floor planning, timing, and area. Table 2.1 summarizes the need and usage for each model.

One of the key concerns in present-day technology is the piracy of IP or core designs. With unprotected models, it is easy to reverse engineer the design, develop an improved design, learn the trade secrets, and pirate the whole design. To restrict piracy and reverse engineering, many of the models are delivered in encrypted form. The most commonly used method is to create a top-level module and instantiate the core model inside it. Thus, the top-level module behaves as a wrapper (shell) and hides the whole netlist, floor planning, and timing of the core. This wrapper uses a compiled version

Table 2.1
Summary of Core Models and Their Usage

Model Type	Development Environment	Need	Usage
ISA	C, C++	Microprocessor based designs, hw/sw cosimulation	High-speed simulation, application run
Behavioral	C, C++, HDL	Nonmicroprocessor designs	High-speed simulation, application run
Bus functional	C, C++, HDL	System simulation, internal behavior of the core	Simulation of bus protocols and transactions
Fully functional	HDL	System verification	Simulation of cycle-by-cycle behavior
Emulation	Synthesized HDL	High-speed system verification	Simulation of cycle-by-cycle behavior
Timing	Stamp, Synopsis.do, SDF	Required by firm and hard cores	Timing verification
Floor plan/area	LEF format	Required by hard cores only	SoC-level integration and physical design

of the simulation model rather than the source code and, hence, it also provides security against reverse engineering of the simulation model.

2.3   Design Process for Soft and Firm Cores

Regardless of whether the core is soft, firm, or hard, the above-mentioned design guidelines are necessary because cores are designed for reuse. The soft and firm cores are productized in RTL form and, hence, they are flexible and easy to reuse. However, because the physical design is not fixed, their area, power, and performance are not optimized.

2.3.1   Design Flow

Soft and firm cores should be designed with a conventional EDA RTL-synthesis flow. Figure 2.5 shows such a flow. In the initial phase, while the core specs are defined, core functionality is continuously modified and partitioned into sub-blocks for which functional specs are developed. Based on

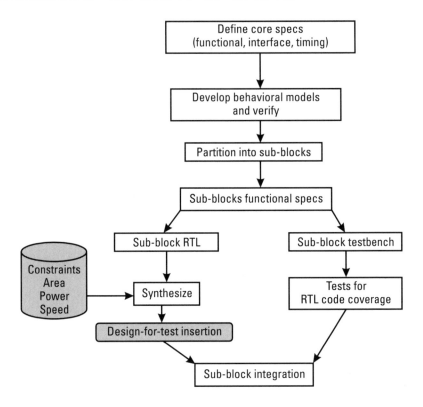

Figure 2.5 RTL synthesis-based design process for soft and firm cores. Shaded blocks represent additional considerations required by firm cores.

these partitioned sub-blocks, RTL code is developed together with synthesis scripts. Timing analysis, area, and power estimations are revised and testbenches are developed to verify RTL. During integration of sub-blocks into core-level design, a top-level netlist is created and used to perform functional test and synthesis.

Because of the reusability requirement, multiple configuration tests should be developed and run. These configuration tests vary significantly depending on whether a soft or firm core is being tested. In general, the synthesis script of firm cores provides a netlist with a target performance and area. Because the netlist of firm cores under this synthesis script is fixed, the testbench for gate-level simulation, the timing model, and the power analysis model can be developed. In the majority of cases, design-for-test methodology (scan, BIST, Iddq) is also considered in the development of firm cores, and fault-grading analysis is done on gate-level netlists. For firm cores, the physical design requirements are considered as the sub-blocks are developed. These requirements consist of interconnects, testbench, overall timing, and cell library constraints.

2.3.2 Development Process for Soft/Firm Cores

At every design step in the core development process, design specifications are needed. General design specifications include the following:

1. Functional requirements to specify the purpose and operation of the core.

2. Physical requirements to specify packaging, die area, power, technology libraries, and so on.

3. Design requirements to specify the architecture and block diagrams with data flow.

4. Interface requirements to specify signal names and functions, timing diagrams, and DC/AC parameters.

5. Test and debug requirements to specify manufacturing testing, design-for-test methodology, test vector generation method, fault grading, and so on.

6. Software requirements to specify software drivers and models for hardware blocks that are visible to software such as general-purpose registers of a microprocessor.

2.3.2.1 Top-Level Design Specifications

The first step in a core design process is to refine the functional specs so that they can be partitioned into self-contained sub-blocks. The general objective is for each sub-block to be able to be designed without interference/dependency of other blocks, as well as coded and verified by a single designer. Sub-block interfaces should be very clearly defined so that assembly of sub-blocks can be conflict free.

In many cases, a behavioral model is used as an executable specification for the core. This model is generally used in the testbench development; for firm cores it is a key simulation model. A behavioral model is essential for a core that has a high algorithmic content such as that needed for MPEG or 2D/3D graphics. For a state machine-dominated core or for a core with little algorithmic content, an RTL model can provide the equivalent abstraction description and simulation performance. In addition to behavioral/RTL simulation models, a testbench with self-checking for output responses is also required at the top level to describe the bus-functional models for surrounding subsystems.

2.3.2.2 Sub-Block Specifications

Sub-block specification starts with the partitioning of the top-level functional model. Creating detailed specifications for sub-blocks allows for efficient RTL coding. EDA tools (memory compilers and module compilers) can be used to generate RTL code if the sub-block consists of structured components such as RAMs, ROMs, FIFOs, and so on. Before RTL coding begins, timing constraints, and power and area specifications are also required. These constraints at block level should be derived from the core-level functional specs.

Along with RTL coding, testbenches are developed to verify the basic functionality of the blocks. At this stage, low-effort first-pass synthesis is done to determine if timing, area, and power constraints can be met. As timing, area, and power are optimized, a synthesis script is developed that is used for sub-block synthesis during integration.

2.3.2.3 Integration of Sub-Blocks

Once the design for the sub-blocks is completed, they are integrated into one design and tested as part of the core. During the initial phase of integration, some mismatches may occur at the interfaces of the sub-blocks. These mismatches can be checked by the core-level RTL model that instantiates sub-blocks and connects them.

Functional tests should be developed using a sufficiently large number of configurations of the core. This ensures the robustness of the final design. When the parameterized core is finalized, it is helpful to provide a set of scripts for different configurations and constraints of the core. Some provisions must be made in timing constraints to account for design-for-test insertion such as scan. Also, a robust power analysis must be done on various configurations of the core.

2.3.3 RTL Guidelines

Good RTL coding is a key to the success of soft/firm cores. Both portability and reusability of the core are determined by the RTL coding style. It also determines the area and performance of the core after synthesis. Therefore, RTL coding guidelines should be developed and strictly enforced in the development of soft/firm cores. The basic principle behind these guidelines should be to develop RTL code that is simple, easy to understand, structured, uses simple constructs and consistent naming conventions, and is easy to verify and synthesize. Some books on Verilog and VHDL are useful in understanding the pros/cons of a specific type of coding style [10–13]. Some basic guidelines are given in Appendix A; these guidelines should be used only as a reference, it is recommended that each design team develop their own guidelines.

2.3.4 Soft/Firm Cores Productization

Productization means the creation and collection of all deliverable items in one package. In general for soft and firm cores, deliverables include RTL code of the core, functional testbenches and test vector files, installation and synthesis scripts, and documentation describing core functionality, characteristics, and simulation results. (Firm cores also required gate-level netlist, description of the technology library, timing model, area, and power estimates.) Because many of the documents created during various development phases are not suitable for customer release, a user manual and data book are also required. This data book should contain core characteristics with sufficient description of design and simulation environments.

As a general rule, prototype silicon should be developed for firm cores and should be made available to the user. Although this prototype silicon results in additional cost, it permits accurate and predictable characterization of the core. Accurate core parameters are extremely valuable in reuse and simplify SoC-level integration.

2.4 Design Process for Hard Cores

The design process for hard cores is quite different from that of soft cores. One major difference is that physical design is required for hard cores and both area and timing are optimized for target technology. Also, hard cores are delivered in a layout-level database (GDSII) and, hence, productization of hard cores is also significantly difficult compared to that of soft cores. In some sense, the design process for hard cores is the same as that for a traditional ASIC design process. Hence, many issues of the traditional ASIC design process [14–16] are applicable to hard cores.

2.4.1 Unique Design Issues in Hard Cores

In addition to the general design issues discussed in Section 2.2, some unique issues are related to the development of hard cores. Most of these issues are related to physical design.

2.4.1.1 Clock and Reset

Hard cores require implementation of clock and reset. This implementation should be independent of SoC clock and reset because SoC-level information is not available at the time of core design. Therefore, to make it self-sufficient, clock and reset in hard cores require buffering and minimum wire loading. Also, a buffered and correctly aligned hard core clock is required to be available on an output pin of the core; this is used for synchronization with other SoC-level on-chip clocks. In general, all the items discussed in Sections 2.2.4 and 2.2.5 should be followed for clock and reset signals.

2.4.1.2 Porosity, Pin Placement, and Aspect Ratio

During SoC-level integration, it is often desirable to route over a core or through a core. To permit such routing, a hard core should have some porosity, that is, some routing channels through the core should be made available. Another possibility is to limit the number of metal layers in the core to one or two less than the maximum allowable by the process. The deliverables for the core should include a blockage map to identify the areas where SoC-level routing may cause errors due to crosstalk or other forms of interaction.

Similar to porosity, pin placement and pin ordering of a core can have a substantial impact on the SoC-level floor plan and routing. As a rule of thumb, all bus signals including external enable are connected to adjacent pin locations; input clock and reset signals are also made available as outputs. In general, large logic cores are placed on one corner of the SoC. Thus, Vdd/Gnd pins should be placed on one or, at most, two sides rather than distributing them along all four sides. This rule is tricky for signal pins. However, the signals that will remain primary I/Os at the SoC level, such as USB and PCI bus, should be placed on one side. Inside the core, common Vdd/Gnd wires should be shorted as rings to minimize voltage spikes and to stabilize internal power/ground.

Another item that can have a serious impact on SoC floor plan and routing is the aspect ratio of the hard core. As much as possible, the aspect ratios should be kept close to 1:1 or 1:2. These aspect ratios are commonly accepted and have minimal impact on SoC-level floor plan.

2.4.1.3 Custom Circuits

Sometimes hard cores contain custom circuit blocks because of performance and area requirements. Because implementation of these circuits is not done through RTL synthesis-based flow, these circuits require schematic entry into the physical design database as well as an RTL model that can be integrated into the core-level functional model. These circuits are generally simulated at transistor level using Spice; hence, an additional timing model is also required for integration into the core-level timing model. In most cases, the characteristics of these circuits are highly sensitive to technology parameters; therefore, good documentation is required to describe the functionality and implementation of these circuits. The documentation with core release should also list these circuits with descriptions of their high-level functionality.

2.4.1.4 Test

Design-for-test (DFT) and debug test structures are mandatory for hard cores but not for soft and firm cores. Thus, core-level DFT implementation requires that it create minimal constraints during SoC integration. A discussion of this process is skipped here because detailed discussions on test issues and solutions are given in Chapters 6 to 10.

2.4.2 Development Process for Hard Cores

A hard core may contain some custom circuits and some synthesized blocks. For synthesized blocks, a design flow such as that given in Figure 2.5 should be followed, while a custom circuit can be simulated at the transistor level, and the design database should have full schematics. Using the RTL model of custom circuits and RTL of synthesized blocks, an RTL model of the full core should be developed. This model should go through an iterative synthesis flow to obtain area, power, and timing within an agreed-upon range (this range can be 10% to 20% of target goals).

During this iteration full design validation should be done for synthesized blocks as well as for custom circuits. The gate-level netlist with area, power, and timing within 10% to 20% of target should be used for physical design. The final timing should be optimized using extracted RC values from the layout-level database. The layout database should be LVS (layout versus schematic) and DRC (design rule checker) clean for a particular technology deck. Finally, various models (functional, bus-model, simulation, floor plan, timing, area, power, and test) should be generated for release. A simplified version of such flow is shown in Figure 2.6.

At the present time, the common situation is that the silicon vendor of the SoC chip is also the provider of hard cores (either in-house developed cores or certified third-party cores). For certified cores, the silicon vendor licenses a hard core, develops various models (with the help of the core provider), and validates the core design and its models within in-house design

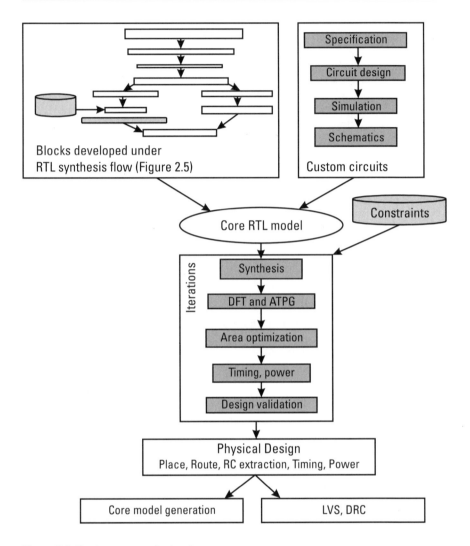

Figure 2.6 Design process for hard cores.

flow before including it in the core library. In the majority of cases, this validation also includes the silicon prototype. Thus, the SoC designer gets the GDSII file along with the timing, power, area, and test models of the hard core.

Hard cores also require much stringent documentation compared to soft cores. This additional documentation (relative to soft cores) includes footprint (pin placement), size of the core in specific technology, detailed timing data sheets, routing and porosity restrictions, Vdd/Gnd and interconnect rules, clock and reset distribution rules, and timing specs.

2.5 Sign-Off Checklist and Deliverables

One purpose of a sign-off checklist is to ensure that certain checks were made during design, simulation, and verification so that the final files meet certain criteria. Another objective of the checklist is to ensure that all necessary design, simulation, and verification files have been created and that installation scripts and required documentation have been developed. These files, scripts, and documentation form the deliverables.

2.5.1 Sign-Off Checklist

The sign-off checklist should include procedures for design checks as well as procedures for database integrity. For design, a check for the following rules is recommended (this list is suitable for hard cores; soft cores will require a subset of this list):

- Completely synchronous design;
- No latches in random logic;
- No multicycle paths;
- No direct combinational paths from inputs to outputs;
- Resynchronization at clock boundary;
- Resynchronization of all asynchronous set/reset/clear signals;
- Synchronized write/read at memory boundary;
- Memory design and placement rule checks;
- Analog/mixed-signal circuits design and placement rule checks;
- Guard bands for memory and analog/mixed-signal circuits;
- Synchronization and protocol verifications for on-chip buses;
- Load balancing in clock tree;
- Isolated clock domains;
- Buffered clocks at the block boundary;
- Clock skew within specified margin;
- Registered block inputs/outputs;
- No combinational feedback loops;

- No internal tri-states;
- No reconvergent logic;
- Static timing analysis done;
- Electromigration rules check;
- No DRC violations;
- LVS and DRC checks for custom circuits;
- RTL and structural simulation match;
- RTL code coverage;
- Gate-level simulation done;
- Fault grading and simulation done;
- Fault coverage;
- SDF (standard delay format) back-annotated timing;
- Functional simulation done;
- DFT rules (such as scan rules) check is done;
- Timing, synthesis, test, design shell files generated.

2.5.2   Soft Core Deliverables

Soft core deliverables are significantly less stringent than hard core deliverables and include the following:

- Synthesizable Verilog/VHDL;
- Example synthesis script;
- RTL compiled module;
- Structural compiled module;
- Design, timing, and synthesis shells;
- Functional simulation testbench;
- Installation script;
- Bus functional models and monitors used in testbenches;
- Testbenches with sample verification tests;
- Cycle-based simulation or emulation models;
- Bus functional models;
- Application note that describes signal slew rate at the inputs, clock skew tolerance, output-loading range, and test methodology.

2.5.3 Hard Core Deliverables

The deliverables for hard cores consist primarily of the models and documentation for the core integrator to design and verify the core in SoC environment. Deliverables include the following:

- Installation scripts;
- ISA or behavioral model of the core;
- Bus functional and fully functional models for the core;
- Cycle-based emulation model (on request);
- Floor planning, timing, and synthesis models;
- Functional simulation testbench;
- Bus functional models and monitors used in testbenches;
- Testbenches with verification tests;
- Manufacturing tests;
- GDSII with technology file (Dracula deck);
- Installation script;
- Application note that describes timing at I/Os, signal slew rate, clock distribution and skew tolerance, power, timing data sheet, area, floor plan, porosity and footprint, and technology specifications.

2.6 System Integration

The key issues in integrating the core into final SoC include logical design, synthesis, physical design, and chip-level verification.

2.6.1 Designing With Hard Cores

Developing a chip using hard cores from external sources such as IP vendors carries certain issues such as from which source to acquire, deign and verification of interfaces between the cores and the rest of the chip, functional and timing verification of the chip, and physical design of the chip.

The most difficult tasks are related to verification. The verification of different aspects such as application-based verification, gate-level verification, and so on requires significant effort. The most important task of SoC design is to verify functionality and timing (performance) at the system level.

Normally, the SoC-level validation effort is about 60% to 75% of the total design effort. Because of the importance of this topic, a detailed discussion of validation is given in Chapter 4.

Various items need to be considered in core selection. These include the quality of the documentation, robustness/completeness of the validation environment that comes with the core, completeness and support for the design environment, and so on. Hard cores generally require that the design be silicon proven with predictable parameters and that the physical design limitations such as routing blockage and porosity of the core are clearly identified.

From the physical design's point of view, distribution of clock, Vdd/Gnd, and signal routing is important for hard cores. The delays in the core must be compatible with the clock timing and clock skew of the rest of the chip since the hard core has its own internal clock tree. Because a hard core would limit or prevent the routing of signals, the placement of the core in the chip can be critical in achieving routability and timing of the chip. The requirements of the power and ground signals and switching characteristics must also be met because they could affect the placement and route.

2.6.2 Designing With Soft Cores

Some of the issues in SoC designs that use soft cores from external sources are same as those for hard cores. These include the quality of the documentation and robustness/completeness of the verification environment that comes with the core.

The core and related files including the complete design verification environment should be installed in the design environment that looks like the core development environment.

Many soft cores are configurable using parameters and the user can set them to generate complete RTL. After RTL generation, the core can be instantiated at the top-level design. The main issue in this process is the correctness of interfaces between the core and the rest of the system. Finally, even if the core provider has verified that the core meets the timing on multiple cell libraries and configurations, the SoC designer should still verify it using target technology library.

2.6.3 System Verification

Along with the SoC specification development, SoC-level behavioral models are developed so that the designer can create testbenches for the verification of the system without waiting for the silicon or a hardware prototype. Therefore, a good set of test suites and test cases are needed, preferably with actual software applications by the time RTL and functional models for the entire chip are assembled. Efficient system-level verification depends on the quality of test and verification plans, quality and completeness of testbenches and the abstraction level of various models, EDA tools and environment, and the robustness of the core.

The system-level verification strategy is based on the design hierarchy. First the leaf-level blocks (at core level) are checked for correctness in a stand-alone manner. Then the interfaces between the cores are verified in terms of transaction types and data contents. After verification of bus functional models, actual software application or an equivalent testbench should be run on the fully assembled chip. This is generally a hardware–software cosimulation. This could be followed by a hardware prototype either in ASIC form or a rapid prototype using FPGAs. Because of the importance of the topic, system verification is discussed in detail in Chapter 4.

References

[1] Keating, M., and P. Bricaud, *Reuse Methodology Manual,* Norwell, MA: Kluwer Academic Publishers, 1998.

[2] International Technology Roadmap for Semiconductors (ITRS), Chapter on Design, Austin, TX: Sematech, Inc., 1999.

[3] Gajski, D., et al., *Specification and Design of Embedded Systems,* Englewood Cliffs, NJ: Prentice Hall, 1994.

[4] Milne, G., *Formal Specification and Verification of Digital Systems,* New York: McGraw-Hill, 1994.

[5] Chrysalis Design Verifier and Design Insight application notes.

[6] VSPEC web page, http://www.ececs.uc.edu/~pbaraona/vspec/.

[7] International Technology Roadmap for Semiconductors (ITRS), Austin, TX: Semtech, Inc., 1999.

[8] Micheli, G. D., *Synthesis and Optimization of Digital Circuits,* New York: McGraw-Hill, 1994.

[9] Knapp, D. W., *Behavioral Synthesis: Digital System Design Using the Synopsys Behavioral Compiler,* Englewood Cliffs, NJ: Prentice Hall, 1996.

[10] Sternheim, E., R. Singh, and Y. Trivedi, *Digital Design with Verilog HDL,* Automata Publishing, 1990.

[11]   Palnitkar, S., *Verilog HDL: A Guide to Digital Design and Synthesis,* Englewood Cliffs, NJ: Prentice Hall, 1996.

[12]   Armstrong, J. R., and F. G. Gray, *Structured Logic Design with VHDL,* Englewood Cliffs, NJ: Prentice Hall, 1993.

[13]   IEEE Standard 1076-1987, *IEEE Standard VHDL Language Reference Manual.*

[14]   Preas, B., and M. Lorenzetti (Eds.), *Physical Design Automation of VLSI Systems,* New York: Benjamin/Cummings Publishing Company, 1988.

[15]   Smith, M. J. S., *Application Specific Integrated Circuits,* Reading, MA: Addison Wesley, 1997.

[16]   Sherwani, N. A., *Algorithms for VLSI Physical Design Automation,* Norwell, MA: Kluwer Academic Publishers, 1993.

3

Design Methodology for Memory and Analog Cores

Similar to the logic cores, design-for-reuse is absolutely necessary for both memories and analog circuits (some key analog circuits used in SoC are DACs, ADCs, and PLLs). As mentioned in Chapter 2, both memories and analog circuits are extremely sensitive to noise and technology parameters. Hence, in almost all the cases, hard cores or custom-designed memories and analog circuits are used. Therefore, design-for-reuse for memories and analog circuits require all of the items described in Chapter 2 for digital logic cores plus many additional rules and checks. In this chapter, we first describe embedded memories and then items that are specific to analog circuits.

3.1 Why Large Embedded Memories

In the present-day SoC, approximately 50% to 60% of the SoC area is occupied by memories. Even in the modern microprocessors, more than 30% of the chip area is occupied by embedded cache. SoCs contain multiple SRAMs, multiple ROMs, large DRAMs, and flash memory blocks. In 1999, DRAMs as large as 16 Mbits and flash memory blocks as big as 4 Mbits have been used in SoC. Another growing trend is that both large DRAM and large flash memories are embedded in SoC. In 1999, 256-Kbits flash memory combined with 1-Mbits DRAM have been embedded in SoCs. According to the 1999 International Technology Roadmap for Semiconductors (ITRS), by 2005, in various applications 512-Mbits DRAM or 256-Mbits flash or 16-Mbits flash combined with 32-Mbits DRAM will be used [1].

The motivations of large embedded memories include:

1. Significant reduction in cost and size by integration of memory on the chip rather than using multiple devices on a board.
2. On-chip memory interface, thus replacing large off-chip drivers with smaller on-chip drivers. This helps reduce the capacitive load, power, heat, and length of wire required while achieving higher speeds.
3. Elimination of pad limitations of off-chip modules and using a larger word width that gives higher performance to the overall system.

The major challenge in the integration of large memory with logic is that it adds significant complexity to the fabrication process. It increases mask counts, which affects cost and memory density and therefore impacts total capacity, timing of peripheral circuits, and overall system performance. If the integrated process is optimized for logic transistors to obtain fast logic, than the high saturation current prohibits a conventional one-transistor (1T) DRAM cell. On the other hand, if the integrated process is optimized for DRAM with very low leakage current, then the performance (switching speed) of the logic transistor suffers.

To integrate large DRAMs into the process optimized for logic, some manufacturers have used three-transistor (3T) DRAM cells; however, this results in a larger area, which limits the integration benefits. In recent years, manufacturers have developed processes that allow two different types of gate oxides optimized for DRAM and logic transistors. Such processes are generally known as *dual-gate processes*. In a dual-gate process, logic and memory are fabricated in different parts of the chip, while each uses its own set of technology parameters. As an example, Table 3.1 illustrates the comparative parameters when the process is optimized for logic versus when it is optimized for DRAM [2]. The cross sections when the process is optimized for performance and DRAM density are shown in Figure 3.1 [2]. As seen from Table 3.1, the cell area (and hence, chip area) and mask count (hence, manufacturing cost) are significantly affected based on whether the process is optimized for logic or DRAM.

Table 3.2 illustrates specific parameters of DRAM cell [2]. The values of current and source-drain sheet resistance clearly identify the reason why performance of 1-transistor cell is lower in Table 3.1. This process complexity is further complicated when flash memory is integrated. Besides dual-gate process, flash memory also requires double poly-silicon layers.

Table 3.1

Memory Cell Comparison in 0.18 μm Merged Logic-DRAM Technology with Four-Level Metal
(From [2], © IEEE 1998. Reproduced with permission)

	Technology Optimized for Logic				Technology Optimized for DRAM			
	1T	3T	4T	6T	1T	3T	4T	6T
Cell area (μ^2)	0.99	2.86	3.80	5.51	0.68	1.89	2.52	3.65
Mask count	25	21	21	19	28	21	21	20
Performance (MHz)	200–250	300	400	400–500	200–250	300	400	400–500

Table 3.2

Memory Cell Parameters in 0.18-μm Technology, Nominal Power Supply 1.8V at 125°C
(From [2], © IEEE 1998. Reproduced with permission)

Cell Type	1T	3T	4T	6T
Access transistor nMOS I_{ON} (mA/μm)	0.05–0.1	0.2–0.3	0.2–0.3	0.5–0.55
Access transistor nMOS I_{OFF} (pA/μm)	0.005–0.05	5–10	5–10	1000–1200
Source-drain sheet resistance nMOS (Ω/sq)	3000–4500	9–15	9–15	9–15
Gate sheet resistance (Ω/sq)	12–50	12–15	12–15	12–15
Source-drain contact resistance nMOS (Ω/contact)	3K–10K	10–30	10–30	10–30
Storage capacitance (fF/cell)	25–35	10–15	10–15	NA
Storage capacitor leakage (pA/cell)	0.01	0.01	0.01	NA
Storage capacitor breakdown (V)	1.5	1.5	1.5	NA

To simplify design complexity resulting from the use of two sets of parameters and the existing memory design technology, memory manufacturers and fabs have developed DRAM and flash memory cores and provided them to the SoC designers. Still, during the simulation, engineers are required to work with two sets of parameters.

3.2 Design Methodology for Embedded Memories

Before a large memory core is productized or included in the library (for example, a multi-megabit DRAM or flash), a test chip is developed for full

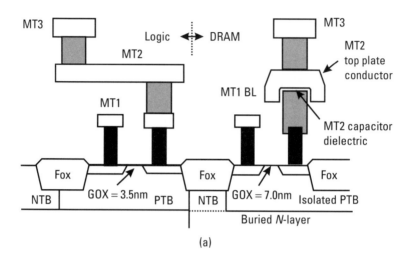

(a)

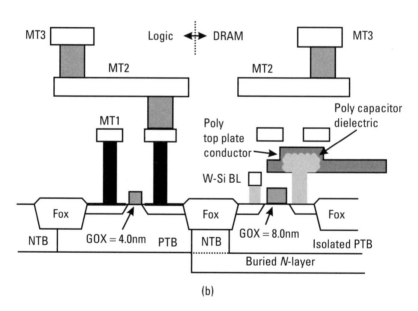

(b)

Figure 3.1 Process cross section of merged logic-DRAM technologies: process optimized (a) for performance and (b) for DRAM density. (From [2], © IEEE 1998. Reproduced with permission.)

characterization. For smaller memories that are designed by memory compiler, extensive SPICE-level simulations are conducted to identify any potential problem and to optimize various characteristics.

3.2.1 Circuit Techniques

The basic structures of SRAM, DRAM, and flash cells are shown in Figure 3.2, while the simple write circuit and sense amplifiers are shown in Figure 3.3. In various applications in SoC, multiport, content addressable, and multi-buffered RAMs are commonly used; the cell structures for these memories are shown in Figure 3.4. These various circuits have different design optimization requirements. For example, the main optimization criteria for the storage cell is area, while the address decoders and sense amplifiers are optimized for higher speed and lower noise. These elements are discussed in separate subsections.

3.2.1.1 Sense Amplifiers

Besides the storage cell, sense amplifiers are the key circuits that are either fully characterized through a test chip or extensively simulated at the SPICE level. Various amplifier parameters are described.

An amplifier's gross functional parameters are given as follows:

1. *Supply currents:* The current source/sink by the amplifier power supplies.

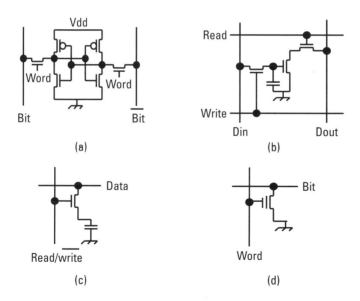

Figure 3.2 Structure of memory cells: (a) six-transistor SRAM cell; (b) three-transistor DRAM cell; (c) one-transistor DRAM cell; (d) flash cell.

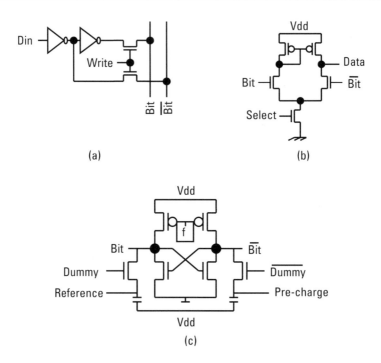

Figure 3.3 Memory circuit elements: (a) write circuit; (b) differential SRAM sense amplifier; (c) differential DRAM sense amplifier.

2. *Output voltage swing (V_{OP}):* The maximum output voltage swing that can be achieved for a specified load without causing voltage limiting.

3. *Closed-loop gain:* The ratio of the output voltage to the input voltage when the amplifier is in a closed-loop configuration.

An amplifier's DC parameters are given as follow:

1. *Input offset voltage (V_{IO}):* The DC voltage that is applied to the input terminals to force the quiescent DC output to its zero (null) voltage. Typically, it ranges from $\pm 10\,\mu$V to ± 10 mV.

2. *Input offset voltage temperature sensitivity (ΔV_{IO}):* The ratio of the change of the input offset voltage to the change of circuit temperature. It is expressed in μV/°C.

3. *Input offset voltage adjustment range [$\Delta V_{IO}(adj+)$, $\Delta V_{IO}(adj-)$]:* The differences between the offset voltage measured with the voltage

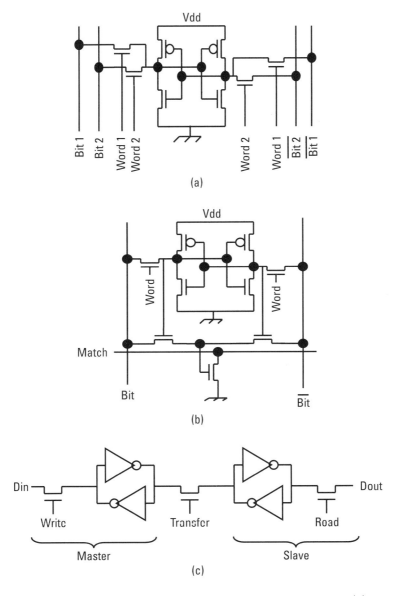

Figure 3.4 Structure of commonly used memories in various applications: (a) two-port memory; (b) content-addressable memory; (c) doubled buffer memory.

adjust terminals open circuited and the offset measured with the maximum positive or negative voltage attainable with the specified adjustment circuit.

4. *Input bias current* $(+I_B, -I_B)$: The currents flowing into the noninverting and inverting terminals individually to force the amplifier output to its zero (null) voltage. Typically, it ranges from 10 pA to 10 μA.

5. *Input offset current* (I_{IO}): The algebraic difference between the two input bias currents.

6. *Input offset current temperature sensitivity* (ΔI_{IO}): The ratio of the change in input offset current to the change of circuit temperature and is usually expressed in pA/°C.

7. *Common mode input voltage range* (V_{CM}): The range of common mode input voltage over which proper functioning of the amplifier is maintained.

8. *Differential mode input voltage range* (V_{DM}): The range of differential mode input voltage over which proper functioning of the amplifier is maintained.

9. *Common mode rejection ratio (CMRR)*: The ratio of the change in input common mode voltage to the resulting change in the input offset voltage. It is given by CMRR $= 20 \log (\Delta V_{CM}/\Delta V_{IO})$ and typically on the order of -100 dB at DC.

10. *Power supply rejection ratio (PSRR)*: The ratio of the change in the input offset voltage to the corresponding change in power supply voltage. It is also on the order of -100 dB.

11. *Open-loop voltage gain* (A_V): The ratio of the change in the output voltage to the differential change in the input voltage.

12. *Output short-circuit current* (I_{OS}): The output current flow when 0V is applied at the output terminal.

13. *Input resistance* (I_R): The resistance as seen by the input terminals.

An amplifier's AC parameters are given as follows:

1. *Small-signal rise time* (t_R): The time taken by the output to rise from 10% to 90% of its steady-state value in response to a specified input pulse.

2. *Settling time* (t_S): The time required by the output to change from some specified voltage level and to settle within a specified band of steady-state values, in response to a specified input.

3. *Slew rate (SR):* The maximum rate of change of output voltage per unit of time in response to input. Typically it is on the order of 100V/μsec.

4. *Transient response overshoot (O_S):* The maximum voltage swing above the output steady-state voltage in response to a specified input.

5. *Overvoltage recovery time:* The settling time after the overshoot, within a specified band.

6. *Unity gain bandwidth:* The frequency at which the open-loop voltage gain is unity.

7. *Gain bandwidth product (GBW):* The frequency at which the open-loop voltage gain drops by 3 dB below its value as measured at DC.

8. *Phase margin:* The margin from 180° at a gain of 0 dB.

9. *Total harmonic distortion (THD):* The sum of all signals created within the amplifier by nonlinear response of its internal forward transfer function. It is measured in decibels, as a ratio of the amplitude of the sum of harmonic signals to the input signal.

10. *Broadband noise (NI_{BB}):* Broadband noise referenced to the input is the true rms noise voltage including all frequency components over a specified bandwidth, measured at the output of the amplifier.

11. *Popcorn noise (NI_{PC}):* Randomly occurring bursts of noise across the broadband range. It is expressed in millivolts peak referenced to the amplifier input.

12. *Input noise voltage density (E_n):* The rms noise voltage in a 1-Hz band centered on a specified frequency. It is typically expressed in nV/$\sqrt{Hz}$ referenced to the amplifier's input.

13. *Input noise current density (I_n):* The rms noise current in a 1-Hz band centered at a specified frequency. It is typically expressed in nA/$\sqrt{Hz}$ referenced to the amplifier input.

14. *Low-frequency input noise density (E_{npp}):* The peak-to-peak noise voltage in the frequency range of 0.1 to 10 Hz.

15. *Signal-to-noise ratio (SNR):* The ratio of the signal to the total noise in a given bandwidth. SNR is measured in decibels as a ratio of the signal amplitude to the sum of noise.

16. *Signal-to-noise and distortion (SINAD):* The ratio of the signal to the sum of noise plus harmonic distortion. The combination of THD and SNR.

3.2.1.2   Floor Planning and Placement Guidelines

Some guidelines related to memories and analog circuit placement, guard banding, on-chip buses, and clock distribution were discussed in Sections 2.2.2 to 2.2.4. These guidelines are very important for large embedded memories. Figure 2.3 also illustrated specific guidelines for memory placement, design of an array with dummy cells, and guard bands.

When an SoC is designed in the merged memory-logic process that contains multi-megabit memory, because of the process complexity some additional placement criteria becomes necessary. For the merged logic-DRAM process, two possibilities are illustrated in Figure 3.5: (1) when the process is optimized for performance and (2) when it is optimized for memory density [2]. Note in Figure 3.5(a) that the 4T cell results in a simple design and provides good performance but requires a large area. On the other hand, in Figure 3.5(b), the 1T cell is used, which allows area optimization, but requires a complex voltage regulator and dual-gate process, yet still provides approximately half the performance of the process of Figure 3.5(a).

3.2.2   Memory Compiler

The majority of memories in present-day SoCs are developed by memory compilers. A number of companies have developed in-house memory compilers; some companies such as Artisan and Virage Logic have also commercialized memory compilers.

These compilers provide a framework that includes physical, logical, and electrical representations of the design database. They are linked with front-end design tools and generate data that is readable with commonly used back-end tools. Based on user-specified size and configuration numbers (number of rows/columns, word size, column multiplexing, and so on), the compiler generates the memory block [3, 4]. The output generally contains Verilog/VHDL simulation models, SPICE netlist, logical and physical LEF models, and GDSII database. From the user's perspective, a generalized flow of memory compilers is illustrated in Figure 3.6.

The format of a few files may vary from one tool to another. Also, some tools may not provide various views of the design and simulation models. For example, memory compilers for various process technologies such as 0.18 and 0.25 μm from TSMC can be licensed from companies such as Artisan.

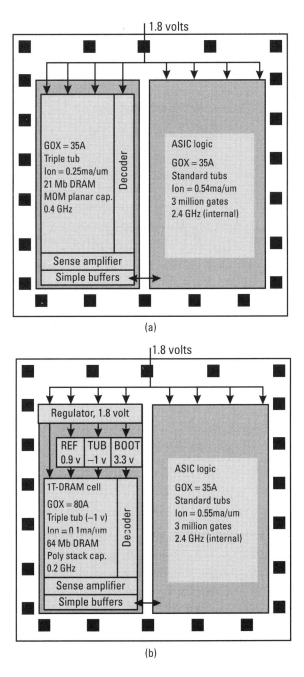

Figure 3.5 Floor planning guidelines for SoC designed in merged logic-DRAM technology: Process optimized (a) for performance and (b) for DRAM density. (From [2], © IEEE 1998. Reproduced with permission.)

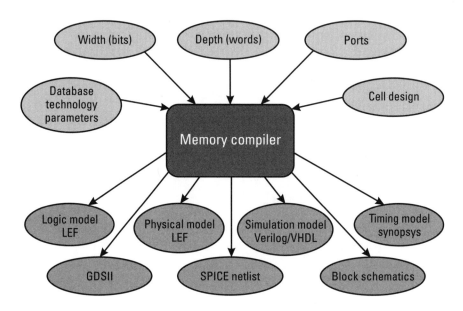

Figure 3.6 General flow of memory compilers.

To support these compilers, standard cell libraries and I/O libraries are also provided. Some example compilers include these:

- High-density single-port SRAM generator;
- High-speed single-port SRAM generator;
- High-speed dual-port SRAM generator;
- High-speed single-port register file generator;
- High-speed two-port register file generator.

These compilers provide PostScript data sheets, ASCII data tables, Verilog and VHDL models, Synopsys Design Compiler models; Prime Time, Motive, Star-DC and Cadence's Central Delay Calculator models, LEF footprint, GDSII layout, and LVS netlist. A user can specify the number of words, word size, word partition size, frequency, drive strength, column multiplexer width, pipeline output, power structure ring width, and metal layer for horizontal and vertical ring layers.

One of the key items in generating high-performance memories from a memory compiler is the transistor sizing. At present the method used in commercial compilers for transistor sizing can be given as follows:

1. Based on memory size and configuration (width and depth), create equations for required transistor width and length. Generally, these are linear equations of the form $Y = mX + c$, where adjusting coefficients m and c affect transistor sizes.

2. Test resulting memory over a range of sizes.

Because memory performance is affected by the transistor sizes, this procedure puts a limit on memory size and configuration; beyond this limit, the compiler becomes unusable. Fortunately, a simple regression-based method can overcome this drawback in transistor sizing [5], as described below.

For a compiler using the min-max range of memory size, four corner cases are defined as follows:

1. Corner a = $(\text{Word}_{min}, \text{bits}_{min})$;
2. Corner b = $(\text{Word}_{min}, \text{bits}_{max})$;
3. Corner c = $(\text{Word}_{max}, \text{bits}_{min})$;
4. Corner d = $(\text{Word}_{max}, \text{bits}_{max})$.

For a memory of width X (number of bits) and depth Y (number of words), an interpolation function that yields transistor width and length from the values determined by corner cases can be given as:

$$F(X,Y) = K_1 + K_2 X + K_3 Y + K_4 XY$$

where the Ks are constants. Thus, the width and length of transistors at corner cases can be given by eight equations (four for width and four for length). As an example, equations for a corner can be given as follows:

$$W_a(X_a, Y_a) = K_1 + K_2 X_a + K_3 Y_a + K_4 X_a Y_a$$
$$L_a(X_a, Y_a) = K_1 + K_2 X_a + K_3 Y_a + K_4 X_a Y_a$$

These equations in the matrix form can be given as follows:

$$[\text{Size}] = [A]\,[K]$$

and thus, the coefficients K_{ij} are given as

$$[K] = [A]^- [\text{Size}]$$

Using this methodology, the design flow with memory compilers is as follows:

1. Create optimized design at each of the corner cases.
2. For every transistor this yields a set of sizes, forming a 4×2 matrix.
3. Store these size matrices in tabular form.
4. Create matrices A, B, C, and D using respective corner values, invert them, and store.
5. Coefficient K_{ij} can be determined for any transistor by $[K] = [A]^-$ [Size].

Now, the width and length of any transistor can be computed for any memory size and configuration by the following equations:

$$W(X,Y) = K_{11} + K_{21}X + K_{31}Y + K_{41}XY$$
$$L(X,Y) = K_{12} + K_{22}X + K_{32}Y + K_{42}XY$$

This transistor sizing allows memories with more predictable behavior even when the memory size is large. It is recommended that SoC designers use such a method with commercial compilers to obtain higher performance, uniform timing, and predictable behavior from SoC memories.

3.2.3   Simulation Models

During the SoC design simulation, Verilog/VHDL models of memories are needed with timing information from various memory operations. The main issue when generating memory models is the inclusion of timing information. The majority of memory compilers provides only a top-level Verilog/VHDL model. Timing of various memory operations (such as read cycle, write cycle) is essential for full chip-level simulation. Memory core vendors provide this information on memory data sheets.

Reference [6] describes a systematic method for transforming timing data from memory data sheets to Verilog/VHDL models. In this method, the timing information is transformed to a Hasse diagram as follows [6]:

1. Label all events indicated on the timing diagram. Let A be the set of all such events.

2. Build the poset on the set $A \times A$ (the Cartesian product of A). An element (a,b) of $A \times A$ is in the poset if there exists a timing link between a and b in the timing diagram.

3. Construct the Hasse diagram from the poset of step 2.

Figure 3.7 illustrates this concept. In the Hasse diagram, each line segment is attached to the timing value taken directly from the data sheet. The transitions occurring on the inputs correspond to the events. As events occur, we move up in the diagram (elapsed time corresponds to the time value associated with line segment). Following the events that occur in correct sequence, we will reach the upper most vertices. An inability to move up in the Hasse diagram reflects an incorrect sequence.

Therefore, converting each vertex into Verilog/VHDL statements while traversing the Hasse diagram transforms timing information into Verilog/VHDL.

The steps to develop device behavior from a set of Hasse diagrams are as follows [6]:

1. Identify the vertices corresponding to changes in inputs. For each such input, designate a variable to hold the value of time of change.

2. For each such vertex, visit the predecessors to develop the timing check. Visit the successor to determine the scheduling action that follows as a result of change.

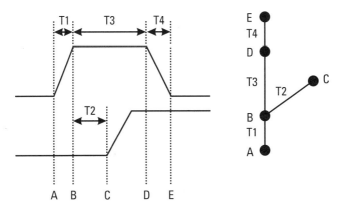

Figure 3.7 Transforming timing data to a Hasse diagram for model generation.

Similar procedures have been used by SoC manufacturers to develop memory models that can be used in full chip simulations. In general, memory core vendors at the present time do not provide such models. In the majority of cases, memory core vendors provide separate data sheets and timing models in specific tool formats (such as Vital and Motive format). Hence, it is recommended that SoC designers use such methods to integrate timing information into the memory simulation model and then integrate the memory simulation model into a full-chip simulation model.

3.3 Specifications of Analog Circuits

While the chip area occupied by the analog circuits varies wildly depending on the application, it is in general hardly 5% of SoC area. The most commonly used analog circuits in SoC are DAC, ADC, PLL, and high-speed I/Os. The primary design issue in analog circuits is the precise specifications of various parameters. For SoC design, the design of an analog circuit must meet the specifications of a significantly large number of parameters to ensure that the analog behavior of these circuits will be within the useful range after manufacturing. Specifications of some commonly used analog circuits are given in separate subsections [7, 8].

3.3.1 Analog-to-Digital Converter

Functional parameters of analog-to-digital converters (ADCs) are shown in Figure 3.8 and described as follows:

1. *Resolution of the ADC* is the basic design specification. It is the ideal number of binary output bits.

2. *Major transitions:* The transition between two adjacent codes that causes all the non-zero LSBs to flip.

3. *Reference voltage (V_{REF}):* An internally or externally supplied voltage that establishes the full-scale voltage range of the ADC.

4. *Full-scale range (FSR):* The maximum (+ve) and minimum(−ve) extremes of input signal (current or voltage) that can be resolved by the ADC as shown in Figure 3.8.

5. *Offset error:* The amount by which the first code transition deviates from the ideal position at an input equivalent to LSB. It is commonly expressed as LSBs, volts, or %FSR, as shown in Figure 3.8.

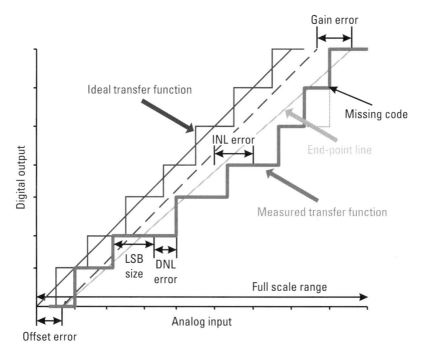

Figure 3.8 DC transfer function and specifications for an ADC.

6. *Gain error:* The deviation of the straight line through the transfer function at the intercept of full scale. It can also be expressed as the deviation in the slope of the ADC transfer characteristic from the ideal gain slope of +1. It is commonly expressed as LSBs, volts, or %FSR as shown in Figure 3.8.

7. *Gain error drift:* The rate of change in gain error with temperature.

8. *LSB size:* The value in volts of the least significant bit resolved by the ADC.

The ADC DC parameters are given as follows:

1. *Supply currents:* The power supply currents are usually measured for the minimum and maximum recommended voltages.

2. *Output logic levels (V_{OL}, V_{OH}):* Output low and high voltage levels on the digital outputs, measured with the appropriate loading I_{OL} and I_{OH}.

3. *Input leakage currents (I_{IH}, I_{IL}):* I_{IH} (I_{IL}) is the input leakage current when applying the maximum V_{IH} (V_{IL}) to the input.

4. *Output high impedance currents (I_{OZL}, I_{OZH}):* Output currents when the output is set to high impedance, for all digital outputs capable of being placed in high impedance.

5. *Output short-circuit current (I_{OS}):* Output current flow when 0V is applied to the output terminal.

6. *Power supply sensitivity ratio (PSSR):* The change in transition voltage for a percentage change in power supply voltage. Generally, PSSR is measured at the first and last transitions.

7. *Differential linearity error (DNL):* The deviation in the code width from the value of 1 LSB.

8. *Monotonicity:* The property that determines that the output of the ADC increases/decreases with increasing/decreasing input voltage.

9. *Integral linearity error (INL):* The deviation of the transfer function from an ideal straight line drawn through the end points of the transfer function, or from the best fit line.

10. *Accuracy:* This includes all static errors and may be given in percent of reading similar to the way voltmeters are specified. This parameter is not tested explicitly, but is implied by all the static errors.

The ADC AC parameters are given as follows:

1. *Input bandwidth:* The analog input frequency at which the spectral power of the fundamental frequency (as determined by the FFT analysis) is reduced by 3 dB.

2. *Conversion time:* The time required for the ADC to convert a single point of an input signal to its digital value. Generally, it is in milliseconds for embedded ADCs, microseconds for successive approximation ADCs, and nanoseconds for flash ADCs.

3. *Conversion rate:* Inverse of the conversion time.

4. *Aperture delay time:* The time required for the ADC to capture a point on an analog signal.

5. *Aperture uncertainty (jitter):* The time variation in aperture time between successive ADC conversions (over a specified number of samples).

6. *Transient response time:* The time required for the converter to achieve a specified accuracy when a one-half-full-scale step function is applied to the analog input.

7. *Overvoltage recovery time:* The amount of time required for the converter to recover to a specified accuracy after an analog input signal of a specified percentage of full scale is reduced to midscale.

8. *Dynamic integral linearity:* The deviation of the transfer function, measured at data rates representative of normal device operation, from an ideal straight line (end points, or best fit).

9. *Dynamic differential linearity:* The DNL (deviation in code width from the ideal value of 1 LSB for adjacent codes) when measured at data rates representative of normal device operation.

10. *Signal-to-noise ratio (SNR):* The ratio of the signal output magnitude to the rms noise magnitude for a given sample rate and input frequency as shown in Figure 3.9.

11. *Effective number of bits (ENOB):* An alternate representation of SNR that equates the distortion and/or noise with an ideal converter with fewer bits. It is a way of relating the SNR to a dynamic equivalent of INL.

12. *Total harmonic distortion (THD):* The ratio of the sum of squares of the rms voltage of the harmonics to the rms voltage of the fundamental frequency.

13. *Signal-to-noise and distortion (SINAD):* The ratio of the signal output magnitude to the sum of rms noise and harmonics.

14. *Two-tone intermodulation distortion (IM):* The ratio of the rms sum of the two distortion components divided by the amplitude of the lower frequency (and usually larger amplitude) component of a two-tone sinusoidal input.

15. *Spurious free dynamic range (SFDR):* The distance in decibels from the fundamental amplitude to the peak spur level, not necessarily limited to harmonic components of the fundamental.

16. *Output/encode rise/fall times:* The time for the waveform to rise/fall between 10% and 90%.

3.3.2   Digital-to-Analog Converter

Digital-to-analog converter (DAC) functional parameters are as follows:

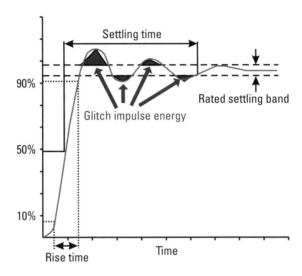

Figure 3.9 Transient response of a DAC showing transient specifications.

1. *Supply currents:* The power supply currents are usually measured for minimum and maximum recommended voltages.

2. *Offset voltage:* The analog output voltage when a null code is applied to the input.

3. *Full-scale voltage:* The analog output voltage when the full-scale code is applied to the input.

4. *Reference voltage (V_{REF}):* An internal or externally provided voltage source that establishes the range of output analog voltages generated by the DAC.

5. *Major transitions:* These are the transitions between codes that cause a carry to flip the least significant nonzero bits and set the next bit.

The DAC DC parameters are as follows:

1. *Full-scale output voltage/current range:* The maximum extremes of output (voltage/current) signal for a DAC.

2. *Offset error:* The difference between the ideal and actual DAC output values to the zero (or null) digital input code.

3. *Gain error:* The difference between the actual and ideal gain, measured between zero and full scale.

4. *LSB size:* The value in volts of the least significant bit of the DAC after compensating for the offset error.

5. *Differential nonlinearity (DNL):* The maximum deviation of an actual analog output step, between adjacent input codes, from the ideal value of 1 LSB based on the gain of the particular DAC.

6. *Monotonicity:* The property that determines the increase/decrease in the output of the DAC with increasing/decreasing input code.

7. *Integral nonlinearity (INL):* The maximum deviation of the analog output from a straight line drawn between the end points or the best fit line, expressed in LSB units.

8. *Accuracy:* An indication of how well a DAC matches a perfect device and includes all the static errors.

9. *Digital input voltages and currents:* These are the V_{IL}, V_{IH}, I_{IL}, and I_{IH} levels for digital input terminals.

10. *Power supply rejection ratio (PSRR):* The change in full-scale analog output voltage of the DAC caused by a deviation of a power supply voltage from the specified level.

The DAC AC parameters are as follows:

1. *Conversion time:* The maximum time taken for the DAC output to reach the output level and settle for the worst case input code change (such as between zero and full scale).

2. *Output settling time:* The time required for the output of a DAC to approach a final value within the limits of a defined error band for a step change in input from high to low or low to high.

3. *Output noise level:* The output noise within a defined bandwidth and with a defined digital input.

4. *Overvoltage recovery time:* The settling time after the overshoot, within a specified band.

5. *Glitch impulse/energy:* The area under the voltage–time curve of a single DAC step until the level has settled down to within the specified error band of the final value.

6. *Dynamic linearity:* The DNL and INL measured at normal device operating rate.

7. *Propagation delay:* The time delay between the input code transition and output settled signal.

8. *Output slew rate:* The maximum rate of change of output per unit of time.

9. *Output rise/fall time:* The time for output to rise/fall between 10% and 90% of its final value.

10. *Total harmonic distortion (THD):* The ratio of the sum of the squares of the rms voltage of the harmonics to the rms voltage of the fundamental.

11. *Signal-to-noise ratio (SNR):* The ratio of the signal output magnitude to the rms noise magnitude.

12. *Signal-to-noise and distortion (SINAD):* The ratio of the signal output magnitude to the sum of rms noise and harmonics.

13. *Intermodulation distortion (IM):* The ratio of the rms sum of the two distortion components to the amplitude of the lower frequency component of the two-tone sine input.

3.3.3. Phase-Locked Loops

The classification of PLL specs is done under open- and closed-loop parameters. In some embedded PLLs, special test modes are also provided to open the VCO feedback loop and provide access to input nodes. Closed-loop parameters of PLLs are given as follows:

1. *Phase/frequency step response:* The transient response to a step in phase/frequency of the input signal.

2. *Pull-in range:* The range within which the PLL will always lock.

3. *Hold range:* The frequency range in which the PLL maintains static lock.

4. *Lock range:* The frequency range during which the PLL locks within one single-beat note between the reference frequency and the output frequency.

5. *Lock time:* The time PLL takes to lock onto the external clock while within the pullout range. It also includes that PLL will not get out of range after this time.

6. *Capture range:* Starting from the unlocked state, the range of frequencies that causes the PLL to lock to the input as the input frequency moves closer to the VCO frequency.

7. *Jitter:* The uncertain time window for the rising edge of the VCO clock, resulting from various noise sources. The maximum offset of the VCO clock from the REF clock over a period of time (e.g., 1 million clocks) gives the long-term jitter. Cycle-to-cycle jitter is obtained by measuring successive cycles. The extremes of the jitter window give the peak-to-peak value, whereas an average statistical value may be obtained by taking the rms value of the jitter over many cycles.

8. *Static phase error:* The allowable skew (error in phase difference) between the VCO clock and REF Clock.

9. *Output frequency range:* The range of output frequencies over which the PLL functions.

10. *Output duty cycle:* The duty cycle of the PLL output clock.

The open-loop parameters of PLL are as follows:

1. *VCO transfer function:* The voltage versus frequency behavior of the VCO. This comprises the following specific parameters: (a) VCO center or reset frequency (f_0), the VCO oscillating frequency at reset; and (b) VCO gain (K_0), the ratio of the variation in VCO angular frequency to the variation in loop filter output signal

2. *Phase detector gain factor (K_d):* The response of the phase detector to the phase lead/lag between the reference and feedback clocks.

3. *Phase transfer function ($H_{j\omega}$):* The amplitude versus frequency transfer function of the loop filter.

4. *3-dB bandwidth (ω_{3-dB}):* The frequency for which the magnitude of $H_{j\omega}$ is 3 dB lower than the DC value.

3.4 High-Speed Circuits

In SoC design, high-speed interface circuits and I/Os are also extremely important. Some example circuits are discussed in separate sections.

3.4.1 Rambus ASIC Cell

Direct Rambus memory technology consists of three main elements: (1) a high bandwidth channel that can transfer data at the rate of 1.6 Gbps (800 MHz), (2) a Rambus interface implemented on both the memory controller and RDRAM devices, and (3) the RDRAM. Electrically, the Rambus channel relies on controlled impedance single terminated transmission lines. These lines carry low-voltage-swing signals. Clock and data always travel in the same direction to virtually eliminate clock to data skew. The interface, called the Rambus ASIC cell (RAC), is available as a library macro-cell from various vendors (IBM, LSI Logic, NEC, TI, Toshiba) to interface the core logic of SoC to the Rambus channel. The RAC consists of mux, demux, TClk, RClk, current control, and test blocks. It typically resides in a portion of the SoC I/O pad ring and provides the basic multiplexing/demultiplexing functions for converting from a byte-serial bus operating at the channel frequency (up to 800 MHz) to the controller's 8-byte-wide bus with a signaling rate up to 200 MHz. This interface also converts from the low-swing voltage levels used by the Rambus channel to ordinary CMOS logic levels internal to SoC. Thus, the RAC manages the electrical and physical interface to the Rambus subsystem.

The channel uses Rambus signaling level (RSL) technology over high-speed, controlled impedance, and matched transmission lines (clock, data, address, and control). The signals use low voltage swings of 800 mV around a Vref of 1.4V, which provides immunity from common mode noise. Dual, odd/even differential input circuits are used to sense the signals. Characteristic impedance terminators at the RDRAM end pull the signals up to the system voltage level (logic 0), and logic 1 is asserted by sinking current using an open-drain NMOS transistor. Synchronous operation is achieved by referencing all commands and data to clock edges, ClockToMaster and ClockFromMaster. Clock and data travel in parallel to minimize skew, and matched transmission lines maintain synchronization. Other specifications include electrical characteristics (RSL voltage and current levels, CMOS voltage and current levels, input and output impedance) and timing characteristics (cycle, rise, fall, setup, hold, delay, pulse widths).

3.4.2   IEEE 1394 Serial Bus (Firewire) PHY Layer

Firewire is a low-cost, high-speed, serial bus architecture specified by the IEEE 1394 standard and is used to connect a wide range of high-performance devices. At the present time, speeds of 100, 200, and 400 Mbps are supported, and a higher speed serial bus (1394.B) to support gigabit speeds is under development. The bus supports both isochronous and asynchronous data transfer protocols. It is based on a layered model (bus management, transaction, link and physical layers) [9, 10].

The physical layer uses two twisted pairs of wires for signaling: one (TPA, TPA*) for data transmission and another (TPB, TPB*) for synchronization. All multiport nodes are implemented with repeater functionality. The interface circuit is shown in Figure 3.10.

Common mode signaling is used for device attachment/detachment detection and speed signaling. The characteristic impedance of the signal pairs is $33 \pm 6\Omega$. Since common mode signaling uses DC signals, there are no reflections. Common mode values are specified as the average voltage on the twisted pair A or B.

Differential signaling is used for arbitration, configuration, and packet transmission. It can occur at speeds of 100, 200, or 400 MHz. It requires elimination of signal by terminating the differential pairs by the characteristic impedance of each signal being (110 Ω). Signal pair attenuation in the cable at 100 MHz is <2.3 dB, at 200 MHz is <3.2 dB, and at 400 MHz is <5.8 dB, so the input receiver tolerates lesser signal strength. A 1394 physical layer core would have specifications that comply with both link layer and cable interfaces including voltage, current, and timing specs.

3.4.3 High-Speed I/O

A wide range of I/O cells are provided by the semiconductor manufacturers and fabs for minimizing I/O bottlenecks suitable for different applications. Commonly offered I/Os includes PCI (peripheral component interconnect) I/Os for PCI buses, PECL (pseudo-ECL) I/Os (single ended and differential), NTL/GTL, HSTL I/Os for low swings, and Hyper-LVDS (low-voltage differential signals) buffers that are compliant with IEEE standards.

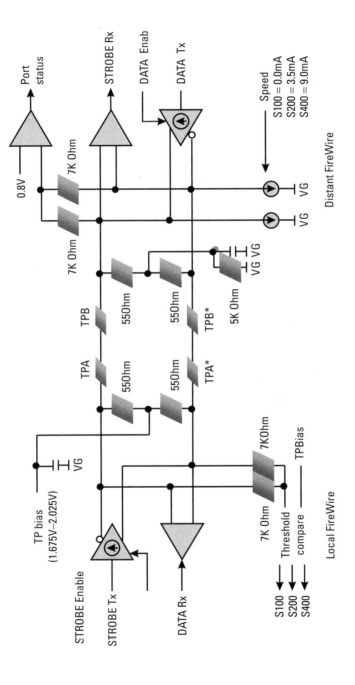

Figure 3.10 Firewire interface circuit. (From [9], © IEEE 1999. Reproduced with permission.)

References

[1]   International Technology Roadmap for Semiconductors (ITRS), Austin, TX: Sematech, Inc., 1999.

[2]   Diodato, P. W., et al., "Merged DRAM-logic in the year 2001," *Proc. IEEE Int. Workshop on Memory Technology, Design and Testing,* 1998, pp. 24–30.

[3]   Tou, J., et al., "A submicrometer CMOS embedded SRAM compiler," *IEEE J. Solid State Circuits,* Vol. 27, No. 3, March 1992, pp. 417-424.

[4]   Lim, H., et al., "A widely configurable EPROM memory compiler for embedded applications," *Proc. IEEE Int. Workshop on Memory Technology, Design and Testing,* 1998, pp. 12–16.

[5]   Donnelly, D., "Memory generator method for sizing transistor in RAM/ROM blocks," *Proc. IEEE Int. Workshop on Memory Technology, Design and Testing,* 1998, pp. 10–11.

[6]   Das, D. V., R. Kumar, and M. Lauria, "Modeling application specific memories," *Proc. IEEE Int. Workshop on Memory Technology, Design and Testing,* 1995, pp. 10–14.

[7]   Mahoney, M., *Tutorial: DSP-Based Testing of Analog and Mixed-Signal Circuits,* Los Alamitos, CA: IEEE Computer Society Press, 1987.

[8]   Demler, M., *High Speed Analog-to-Digital Conversion,* San Diego, CA: Academic Press, 1991.

[9]   Blancha, B., and L. Melatti, "Testing methodology for Firewire," *IEEE Design and Test of Computers,* Vol. 16, No. 3, July-Sep. 1999, pp. 102-111.

[10]   Anderson, D., *Firewire System Architecture,* Reading, MA: Addison Wesley, 1998.

4

Design Validation

Design validation is one of the most important tasks in any system design project. Design validation means establishing that the system does what it was intended to do. It essentially provides confidence in the system's operation. Verification on the other hand means checking against an accepted entity—generally, a higher level of specifications. In IC design, verification is divided into two major tasks: (1) specification verification and (2) implementation verification.

Specification verification is done to verify that during system design, the translation from one abstraction level to the other is correct and the two models match each other. The objective of implementation verification is to find out, within practical limits, whether the system will work after implementation (find out if it works as designed). It is different than validation; the objective of validation is to prove that it indeed works as intended.

Design validation of SoC can be considered as validation of hardware operation, validation of software operation, and validation of the combined system operation. This includes both functionality and timing performance at the system level.

In the early phase of SoC design, along with the specification development and RTL coding, behavioral models are developed so that the testbenches can be created for system simulation. In the early phase, the objective should be to develop a good set of test suites and test cases (including actual software application) by the time RTL and functional models are specified. Efficient validation depends on the quality of test and completeness of testbenches, the abstraction level of various models, EDA tools, and the simulation environment.

The design verification strategy follows design hierarchy. First the leaf-level blocks (at the core level) are checked for correctness in a stand-alone way. In SoC, hardware design consists of dealing with cores and some glue logic (for the sake of uniformity, glue logic can be considered as another core). After functionality checking of these cores, the interfaces between the cores are checked for correctness in terms of transaction types and data content. The next step is to run application software or equivalent testbenches on a full-chip model. Because an application can only be verified by run-time executions of the software code, a hardware–software cosimulation is required. This is followed by emulation and/or a hardware prototype either in ASIC or FPGAs.

4.1   Core-Level Validation

The main difficulty of core-level validation is the development of testbenches that can be reused at the SoC level as well as the validation of functionality when core specifications are parameterized. Testbench reuse is an essential criterion in the development of core-level testbenches because SoC designers need to verify a core's functionality in the context of its final application.

4.1.1   Core Validation Plan

The core design validation plan should be created very early in the design process. Having such a plan early on allows the design team to focus on the area that establishes the base functionality and minimizes redundant efforts. It also provides a formal mechanism for correlating project requirements, and it helps when assessing the completeness of test suites and documenting validation tests and testbenches, a necessary step for reuse. Having a validation plan early on also allows the test environment to be created in parallel with the design.

A general core design flow is shown in Figure 4.1. This flow is essentially the same as the conventional ASIC design flow, following hierarchy. In this flow, the core specs are verified at the behavioral level; this behavioral model is used as a reference model for other abstraction levels. Every time the model is translated to a lower level of abstraction such as RTL, gate level, or layout, the implementation verification as well as the timing verification are done. Although the deliverable model for soft and firm cores is RTL, it is

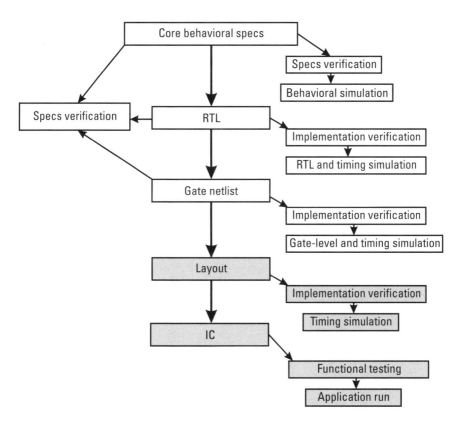

Figure 4.1 Core validation flow. Shaded blocks are applicable only to hard cores.

highly recommended that a gate-level model as well as a layout model be developed and timing simulations performed to validate performance. For hard cores, silicon prototyping is recommended in multiple technologies.

Table 4.1 lists core design at different levels of abstraction and lists the methodologies applicable at each level. Design validation of a core consists of several stages, such as simulation/testing of individual sub-blocks, verification of block interfaces, and validation of the entire core via an application run. The basic types of tests used in this process are compliance testing; corner testing, which includes complex scenarios and min-max cases; random testing; real code testing (application run); and regression testing.

The tools used in this process include event- and cycle-based simulators. Event-driven simulators are good for small size designs and for debugging an error, whereas cycle-based simulators are good for keeping simulation times manageable for large designs. Other useful tools are

Table 4.1

List of Validation Methods at Various Abstraction Levels

Abstraction Level	Validation Methods
Core specs	Simulation in C, C++
Behavioral HDL	System simulators, HDL simulators
Register transfer level	HDL simulators, code coverage, formal verification
Gate level	Formal verification, gate-level simulators, static timing analysis, hardware accelerators, emulators
Physical design	Layout vs schematics (LVS), design rule checker (DRC)

hardware modeling tools that interface between physical chip and software, emulation systems, and hardware prototyping. In general, emulation systems are costly and require long compile times; hardware prototyping is even more costly but it allows for real-time application runs.

4.1.2 Testbenches

Development of a testbench depends largely on the functionality of the core and its intended use. In general, the testbenches can be broadly categorized in the following ways:

1. *Testbench to verify transactions:* This is a very important step for bus controller and interface circuits or whenever a bus functional model is used. It contains two items: (a) the generation of input stimuli for valid transactions with respect to both data and control signals, and (b) response checking.

2. *Testbench to verify instructions:* This verification is particularly important for microprocessor and DSP-type cores. It uses an ISA (instruction set architectural) model for instruction validation. All opcodes should be included in this testbench.

3. *Testbench for random testing:* Random testing is quite useful for discovering obscure bugs. All designs should be exposed to some random testing.

4. *Regression testbench:* This testbench contains a test suite that is developed to identify specific bugs. With the discovery of every new bug, a new test should be included in the regression suite that

is targeted toward the bug. It is also used to ensure that one fix does not create a new error.

5. *Testbench for code coverage:* This test checks for any missed function calls, returns, unused code, wrong branches/returns, incorrect or incomplete loops, and the like. A number of commercial tools can automatically generate tests for code coverage.

6. *Functional testbench:* This is essentially functional testing through an actual application run. Because of the limitation of computing resources, generally, a limited number of application cycles are simulated on the final design.

The importance of these testbenches varies depending on the type of core. For example, a testbench for a processor will definitely consist of a test program based on its instruction set. But the testbench for bus controllers such as USB and PCI will be largely dependent upon the bus functional models, bus monitors, and transaction protocols to apply stimulus and check the simulation results. Similarly, there are significant differences in the sub-block and core-level testbenches. These are separately discussed in Sections 4.1.2.1 and 4.1.2.2.

4.1.2.1 Sub-Block Testbenches

Testbenches for sub-blocks are not very systematic or structured since these are developed by the logic designer; however, some logic designers write good structured tests and those tests are not precluded. A sub-block testbench can be limited to generating a set of input stimuli and to checking the responses at the output ports. In a testbench at sub-block—level, bidirectional ports should be modeled as separate input and output ports. The focus of input stimuli should be the legal transactions at the sub-block level. Thus, input stimuli should include the properties of transactions at the input ports and the constraints of logic. The constraints should include relative timing on data as well as fixed timing of control signals. Checking of output response is generally done manually by the engineer who designed the block.

The stimulus generators should cover corner cases that are most likely to break the design. Most of these tests are deterministic in nature; hence, separately random pattern tests should be developed, because they are useful in identifying obscure bugs. Additionally, code coverage metrics should be used to identify any missing functional calls, return calls, unused codes, branch conditions, statement and path coverage, incorrect and incomplete loops, and so on. At sub-block level 100% code coverage should be achieved before a sub-block is accepted for integration.

4.1.2.2   Core-Level Functional Testbenches

Core-level testbenches are similar to sub-block testbenches, but need to be more sophisticated and well documented because the core testbench is one of the deliverable items when a core is productized. Because the size of output data can be large, manual checking of output response is not suitable; therefore, automated checking is necessary for the output of a core testbench. Note that in a cosimulation environment only a small fraction of the total cycles of application could be simulated. With the testbench based on a bus functional model, a sufficiently larger number of simulation cycles can be generated and the bus monitors can check the output response. However, unless careful attention is paid, these cycles may not be true representatives of the actual application cycles and, hence, may not be useful in finding the actual bugs.

Note also that there are multiple ways to develop a testbench for any given design, hence, a careful evaluation should be done before choosing a testbench development methodology. As an example, to illustrate the trade-offs in the various ways available to develop a functional testbench for the same block, Figure 4.2 illustrates two testbench environments for a USB bus controller interface: (a) using a bus functional model and (b) using an actual application run with hardware–software cosimulation.

For this example, the testbench development is quite easy when using a bus functional model and bus monitors because it can be controlled with a single testbench command file. An actual software application run (in cosimulation) means simulation of actual transactions between the application and the USB controller, and that is not efficient with respect to computational resources. On the other hand, a large number of simulation cycles can be used without difficulty during input stimulus generation for a testbench that uses a bus functional model. Further, the response can be checked automatically by the bus monitors. Hence, it can be concluded that the testbench development using a bus functional model and bus monitors is suitable for this example.

4.1.3   Core-Level Timing Verification

Timing verification is one of the most difficult tasks in the core design process. Static timing analysis and gate-level logic simulation are the most effective methods for accomplishing timing verification at the present time.

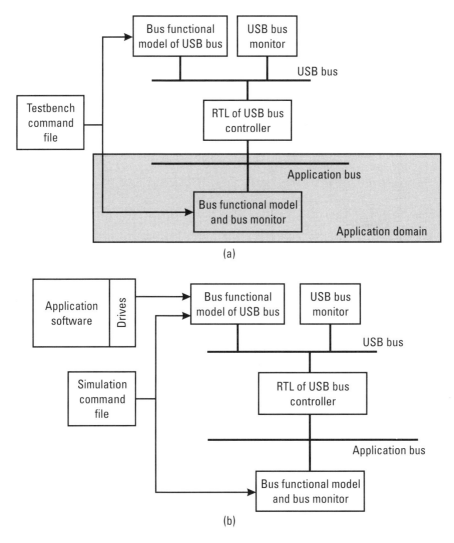

Figure 4.2 Testbench development for USB bus controller using (a) bus functional model and (b) a hardware–software cosimulation.

Static timing analysis is faster but is generally pessimistic because false paths are not properly filtered out. Hence, after synthesis, gate-level simulation should be done although it is slow and computationally intensive. Figure 4.3 illustrates the timing validation flow.

Static timing analysis should be performed on representative netlists of the core that have been synthesized with different technology libraries. It is

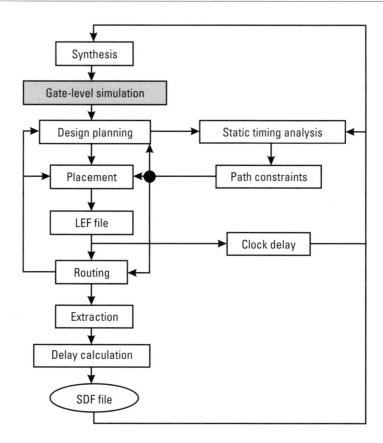

Figure 4.3 Core-level timing validation flow. Shaded box represents recommendation that the gate-level simulation should be performed whenever design size is manageable.

always recommended that a representative (sample) layout be included to verify the timing of the final design even for the soft cores. The primary reason is that the timing can vary significantly depending on the core's synthesis method and layout; such an exercise, however, provides a reference point to the SoC designer. This issue does not arise in hard and firm cores because synthesis script and timing models are included in the deliverables.

The inputs to commercial static timing analysis tools include netlist, technology library cells, timing constraints such as clock period, false paths, waveforms, skews, input, and output delays of the blocks, drive for inputs, loads for outputs, and multicycle paths. Note that the majority of currently available commercial tools consider timing parameters only on the technology library cells; tools that can include delay based on interconnect are still in research. Static timing analysis tools do not require any test vectors. Using parameters as mentioned above, the tool calculates delay through the combinational logic and setup/hold times for registers, and identifies the "time slack" in all paths in the design. The advantage of static timing tools over the logic simulation is their speed and the exhaustive nature of the analysis without the need for test vectors. However, it does not ensure functional correctness of the design and is dependent on the delay model used.

Gate-level simulation is also very useful in timing verification but its limitation is that it takes an excessive amount of time to develop stimulus and to simulate all the timing paths at gate level. Also, the worst case timing scenarios are very difficult and may never be exercised in gate-level simulation because they are difficult to identify manually and automatic methods are not available. The main benefits of gate-level simulation are:

1. Verification that synthesis has generated a correct netlist. Although formal methods can be used to check the correctness of a synthesized netlist in comparison to RTL, gate-level simulation can identify errors that were overlooked in RTL.

2. Verification that the postsynthesis logic insertions such as clock tree, buffers, and scan logic have not changed the functionality.

3. Verification of timing.

4. Verification that the manufacturing tests are correct.

Gate-level simulation is also useful in verifying timing of the asynchronous logic in the core. Note that, as mentioned in Chapter 1, asynchronous logic should be avoided whenever possible. However, if there is some asynchronous logic in the core, full timing gate-level simulations should be targeted for verification. Static timing analysis cannot verify timing of asynchronous logic. The best way to verify timing of a core in all cases is to use static timing and then follow up with gate-level simulation as an additional check.

4.2 Core Interface Verification

In many present SoC designs, on-chip addressing and data buses provide interface and connectivity among the cores. These buses operate by a controller under preset rules such as polling and request/grant protocols. Thus, it is necessary to validate the controller's operation as well as the protocol behavior. Note that the controller itself could be considered to be a core. Hence, the validation methodology described in Section 4.1 is applicable for the bus controller. The key difference in interface validation and core validation is the protocol behavior for data transfer across the core boundary. For this purpose, a limited number of transactions are simulated with various data and control values.

4.2.1 Protocol Verification

To verify the protocol behavior, it is necessary that all transaction types that may take place are clearly defined. Creating a list of all transactions types and developing at least one test for each transaction type is strongly advised. In most SoC designs, not all possible transaction types (supported by the protocol) are allowed. Transaction types are generally restricted to a small number. Thus, identifying all supported transaction types and developing test cases for them is not an overwhelming task. To further automate the testing of protocol behavior using multiple sets of data in any transaction, simple and regular structures such as bus monitors can be used with on-chip buses.

By building bus monitors within the bus controller, both observability and controllability can be enhanced for the verification of transaction protocols. Built-in bus monitors allow control and observation of precise transactions in proper order and are also useful in the debugging process. However, note that the bus functional models are necessary instead of a fully functional model when bus monitors are built within the bus controller.

In the modular design, the behavioral and RTL description of an interface block can be identified separately from the core's internal logic. This separation allows easy monitoring of the interaction among the interface

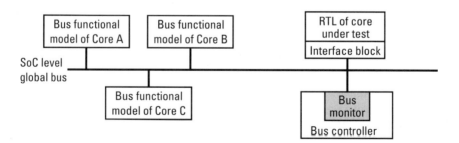

Figure 4.4 Illustration of the use of a bus functional model and bus monitors to verify core interface.

blocks using bus monitors and simulation at behavioral as well as RTL levels. This concept is illustrated in Figure 4.4.

In this method, the bus functional model of every core except the core-under-test replaces the physical netlist of each core. For each interface block, an RTL testbench is created that generates the test sequence using the bus functional models of various cores and a behavioral or RTL model of the core-under-test. The deterministic test data (stimulus) for this testbench should be based on the user's expectations of transaction data. This approach allows the simulation of interfaces and transaction protocols on global buses as well as in point-to-point interfaces when data from one core to another is transferred directly and not through a global bus.

One limitation of this method is that it does not ensure correct behavior for all possible data values and for all sequences that each interface would receive; hence, random data should also be used as part of the stimulus. Use of random data requires special consideration so that core logic is not forced into an illegal state and so that illegal data sequences do not occur at the core's inputs. Thus, either during the random data generation a filter should be used or a checker should be built with the interface that can suppress illegal data.

The response checking is done manually because it is difficult to characterize expected values even when deterministic data are used for testing. Therefore, automatic checking is limited to only generic results such as checking for illegal output transactions and state machine loops.

4.2.2 Gate-Level Simulation

The gate-level netlist of bus interface logic should be verified for both functionality and timing.

Formal verification can be used to verify the correctness of the gate-level netlist. The timing verification flow such as that shown in Figure 4.3 should be used, followed by the gate-level simulation using back-annotated delays in the Verilog/VHDL netlist. Generally, the number of gates in the bus interface logic is small and, hence, gate-level simulation can be accomplished efficiently.

4.3 SoC Design Validation

SoC-level design validation requires a real application run of sufficient length. Thus, fully functional models of all cores and interfaces are needed.

This simulation cannot be done at logic level or even at RTL level because the present-day logic- and RTL-level simulation tools are too slow. For example, the simple booting-up of a real-time operating system (RTOS) at RTL level can take more than 24 hours. Thus, higher levels of abstractions are used that include the ISA model, bus functional model (BFM), and C/C++ models.

At SoC-level validation, generally, RTL models are used for the functional cores, behavioral and ISA models for memory and processor cores, and bus functional models and bus monitors for interface and communications blocks. Most of the time, errors are detected manually using a bus monitor or by the sequence monitor on the communications interfaces. Figure 4.5 illustrates the approximate relative speed of simulation at different levels of abstraction. In Figure 4.5, simulation speeds are illustrated relative to simulation speeds at the logic level, which is generally slower than 10 cycles per second at logic level. The simulation speed at RTL level is generally about 100 cycles per second.

The simulation speed can be improved using cycle-accurate models. In cycle-accurate models, all implementation detail is removed, but enough timing information is kept so that all transactions are accurate at clock

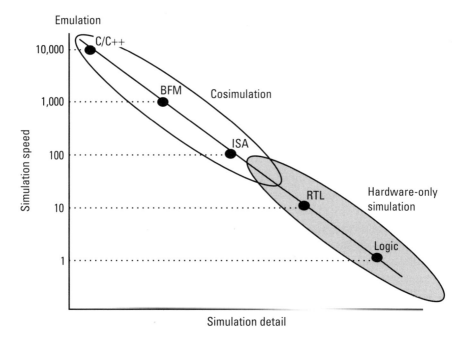

Figure 4.5 Approximate comparison of simulation speed at various abstraction levels.

boundaries. BFM and ISA models further provide one order of magnitude improvement in simulation speed. Using higher level models in C/C++, the simulation can be sped-up to two to three orders of magnitude over the simulation speed at RTL level. More than 10,000 cycles per second, hardware–software cosimulation, can be done and small fragments of real application code can be used for design validation.

4.3.1   Cosimulation

System specs for cosimulation flow are shown in Figure 4.6. In cosimulation, hardware can be modeled in C/C++ and the entire system can be executed like a single C/C++ program. However, it would not be system verification at the implementation level; rather it is a behavioral verification. HDL/RTL description is required for verification because it represents the implementation of hardware components.

Cosimulation requires one or more HDL simulators and a C/C++ platform (compiler, loader, linker, and other pieces from the computer operating system). In cosimulation, two or more simulators including HDL simulators and software simulators are linked. Hence, the communication between different simulators is a key item.

The simplest communication structure for a multisimulator environment is the master/slave behavior. In master/slave operation, the HDL simulator acts as the master and the software simulator behaves as the slave. Slave simulators are invoked using techniques similar to procedure calls and can be implemented by either calling an external procedure from the master simulator or by encapsulating the slave simulator within a procedure call. Most HDL simulators provide a rich set of external procedure calls for invoking C functions. For example, Verilog provides PLI/VPI functions that can be used to enclose C code. VHDL supports foreign attributes within an associated architecture block. This feature allows part of the VHDL code to be written in C. These procedural calls impose some restrictions on the execution of slave simulators, such as requiring that slave simulator functions be executed within a single invocation. This implies that the C program has to be made into a set of C function calls seemingly independent of each other. This scheme becomes too complex to use when there are multiple interaction points with the rest of the hardware model. Also, it is almost impossible to achieve concurrency using individual C function calls because the master/slave combination still provides a single execution process.

The distributed cosimulation model overcomes these restrictions. Distributed cosimulation is based on a communication network protocol that is

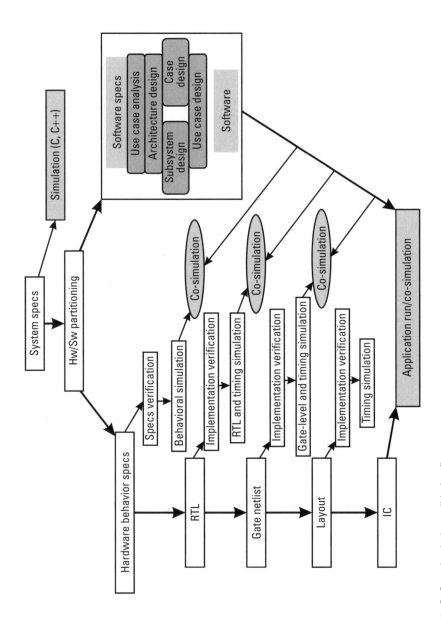

Figure 4.6 SoC cosimulation validation flow.

used as a software bus. Both master/slave and distributed cosimulation models are shown in Figure 4.7.

In distributed cosimulation, each simulator communicates with the cosimulation bus through prespecified procedure calls. The implementation of the cosimulation bus may be based on standard system facilities such as Unix IPC (interprocess communications) or Sockets. It can also be implemented as an ad hoc simulation backplane. In this structure, the HDL simulator and the C program run concurrently as independent software programs (HDL simulators are also C programs). Cosimulation models (C program and the HDL processes) should be independent of the communication mechanism used. There are two advantages to this method:

1. Modularity is achieved because various modules can be designed using different HDL languages.
2. Flexibility is achieved because an HDL model can be replaced by a model at a higher level of abstraction for increased speed, or with a model at a lower level of abstraction for more details as the design progresses.

Figure 4.8 illustrates how debugging can be done in a cosimulation environment. It also shows the flow for generating the executable code.

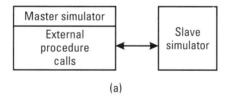

(a)

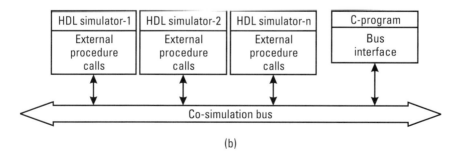

(b)

Figure 4.7 Cosimulation models: (a) master/slave model and (b) distributed simulation model.

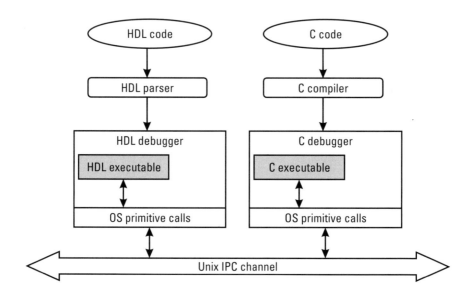

Figure 4.8 Cosimulation debugging model.

Figure 4.8 essentially provides a more detailed view of Figure 4.7 in that it shows the operating system calls of primitive functions in software modules. These calls are a set of C functions available in a library, which allows the controlling, sending, and receiving of data from a Unix/IPC channel. By calling these functions, the C program (software application) can communicate with the IPC channel. The software part is run as a separate process in the host computer operating system, while the executable code is wrapped inside the C debugger.

Similarly, when the C program interacts with a set of hardware modules (by accessing I/O ports of VHDL/Verilog entities), it calls OS primitives to interact with the Unix/IPC channel.

On the hardware side, HDL executable is run by the HDL debugging tool (the HDL simulator can also be executed without the debugger but executing it through the debugging tool is much more useful in cosimulation) and interacts with the IPC channel via the HDL I/O ports. The IPC channel is accessed by the HDL entities through the OS primitive calls, which are C functions wrapped inside the HDL entities to access the IPC channel. (In VHDL, C functions can be invoked by foreign attributes and in Verilog by PLI.) These OS primitive procedures perform the tasks of C-HDL data type conversion, synchronization, and propagation of events between the HDL simulator and the IPC channel.

4.3.2 Emulation

As illustrated in Figure 4.5, the cosimulation speed at the C/C++ level is about 10,000 cycles per seconds. This speed is barely enough to simulate a small fragment of an application. To simulate any reasonable length of real-time application, simulation speeds on the order of a million cycles per second are needed. Furthermore, cosimulation does not verify implementation; it essentially validates the behavioral specs of the system.

These problems are overcome by emulation. Although emulation is still 1 to 2 orders of magnitude slower than the actual silicon, more than 1 million clocks per second can be simulated and, thus, decent size application segments can be run. One major advantage of emulation is that "what if" scenarios of design can be examined and bugs can be fixed. Current emulation technology uses synthesized RTL; hence, it can be viewed as a method for checking functional correctness at the implementation level. Because of the different speed in emulation and actual silicon, emulation is not very useful in detecting timing errors. Other disadvantages of emulation are compilation time after every change and different design flow from the actual silicon implementation flow [1]. In general, emulation technology provides programmable interconnect, fixed-board designs, relatively large gate counts, and special memory and processor support. Emulation can give fast performance if the entire design can be loaded into the emulation engine itself. Performance degrades if a significant portion of the chip or the testbench is loaded onto the host computer.

Present-day emulation systems can be broadly divided into two categories: (1) FPGA-based systems and (2) custom systems. In FPGA-based systems, the whole design netlist is partitioned and mapped onto off-the-shelf FPGAs. Examples of this category are Quickturn's Animator, System Realizer, Mentor Graphics SimExpress, Zycad's Paradigm RP and XP, Apix's System Explorer, and Pro Explorer.

Custom system emulators are based on a unique architecture with numerous special-purpose processors to emulate logic. In these systems, a compiler maps each logic gate onto a special-purpose processor. Examples are Quickturn's Cobalt and Synopsys Arkos.

4.3.3 Hardware Prototypes

Despite the best attempts by engineers to make the first silicon fully functional, only about 80% of designs work correctly when tested at the wafer level, and more than half fail when put into the system for the first time. The primary reason is the lack of system-level testing with a sufficient amount of real application runs. The only means to do it with present-day technology is by using either FPGA/LPGA or ASIC silicon prototypes.

When the design is too large, building a silicon prototype and debugging it in the final system is the best method and is more cost- and time-efficient than the extended verification and cosimulation methods. With a silicon prototype, the same silicon can be used for the first several bugs without a new silicon run. However, deciding when to use silicon is an important decision because producing a silicon prototype is a costly proposition. These factors should be considered:

1. *The diminishing bug rate in verification and cosimulation.* When the basic bugs have been eliminated, an extensive application run will be required to find other bugs. Cosimulation and emulation may not be able to run an application for extensive time periods.

2. *The difficulty of finding bugs.* If finding a bug takes a few orders of magnitude more time than the time required to fix it, then a silicon prototype is very useful because it finds bugs quickly.

3. *The cost (manual effort, time-to-market) of finding bugs.* If the search for bugs with the cosimulation or emulation methods is extremely costly, then silicon prototyping should be considered.

For smaller designs, FPGA/LPGA (field programmable gate-array/laser programmable gate-array) prototypes are adequate. LPGAs allow reprogrammability so the bug fixes can be retested, whereas FPGAs offer higher speeds and higher gate counts than LPGAs. Both FPGA and LPGA lack the gate count capacity and speeds of ASICs, so they are good for smaller blocks or cores, but not suitable for the entire SoC.

Several FPGAs can be used to build a prototype of the entire SoC. In this case, if a bug fix requires the chip to be repartitioned, interconnects between FPGAs will require change, and hence modifications can be complex. A partial solution to this problem is to add custom programmable routing chips to interconnect FPGAs. With programmable routing chips, if the interconnects change, routing can be done under software control.

In a prototype flow, after studying the system's behavior, simulated HDL and/or a C program description is used in cosynthesis. Similar to cosimulation, cosynthesis refers to the simultaneous development of the hardware and software. The purpose of cosynthesis is to produce C code and hardware that will execute on a real architecture. In general, this step consists of mapping the system onto a hardware–software platform that includes a processor to execute the software and a set of ASICs to realize the hardware. The final prototype is generated by using the C code of the software parts and by using logic synthesis, placement, and routing to create the hardware components. Synthesis tools translate the HDL descriptions/models into gate-level netlists that are mapped to FPGAs or ASICs as prototypes.

Reference

[1] Rosenstiel, W., "Prototyping and emulation," in *Hardware/Software Codesign: Principles and Practice* (J. Staunstrup and W. Wolf, Eds.), Norwell, MA: Kluwer Academic Publishers, 1997, pp. 75–112.

5

Core and SoC Design Examples

SoC is in the early development phase and, therefore, every aspect of its design, testing, manufacturing, applications, and even business models is continuously evolving. A large number of companies as well as standards organizations are actively working to resolve various technical issues related to the design of the cores and their use in the SoC.

As mentioned in Chapter 1, commonly used cores in SoC include microprocessor cores, memory cores of various kinds and sizes, application-specific cores such as DSP, bus controllers, and interface cores. The design style and complexity of these cores vary widely. This chapter (or even this book) cannot capture all the different design styles of the various cores. However, a few selected examples are included in this chapter of cores, followed by a few examples of SoC.

5.1 Microprocessor Cores

To say that a core has "high value" means that the design of the core cannot be created easily but is critical to the overall design. Microprocessor cores are one good example of high-value cores because microprocessor/microcontroller design is nontrivial and it is one of the key components of an SoC.

A large number of microprocessor cores are available for embedded applications. Besides the older generation processors, IP providers and ASIC companies have also developed shrink versions of today's popular processors. ARM, MIPS, PowerPC, SPARC, ARC, and MiniRISC are just a few examples of successful highly-valued microprocessor cores. Table 5.1 lists some characteristic parameters for a few selected microprocessors cores. Note that besides speed, power, area, and performance, the architecture of these microprocessor cores also varies in many fundamental ways such as from X86 to RISC to CISC to VLIW. For the SoC designer, the availability of different architectures allows the building of a better configuration for a specific application. To illustrate some sample microprocessors architectures, block diagrams of ARM and AMD's 486DX core are given in Figures 5.1 and 5.2, respectively.

It is worth mentioning that the design of a microprocessor core can be optimized to the specific technology and may thus be available from different companies. In such a case, parameters such as performance, area, and power will change for the base design. To fulfill the needs of SoC designers, large

Table 5.1

Characteristics of Some Popular Microprocessor Cores

Processor		Clock (MHz)	Word Size (bits)	Technology (μm)	Transistor Count	Power (mW/MHz)	Cache (I/D) (Kbyte)
ARM	7TDMI	66	32	(0.6–0.25) (3.3–2.5)	360K	0.6	8 unified
	9TDMI	120–200	32	(0.35–0.25) (3.3–2.0)	570K	0.7	16/16
	SA-110	100–233	32	(0.35) (2)		300–1000	16/8
Intel 960	JA	33	32	(0.8) (3.3)	750	0.5	2/1
	HT	75	32	(0.6) (3.3)	2300	4.5	16/8
MIPS	R4100	40	64	(0.5) (3.3)	450	0.12	2/1
	R4300i	133	64	(0.35) (3.3)	1700	2.2	16/8
	R4650	133	64	(0.6) (3.3)	1050	3.8	8/8
Power PC	401GF	25–100	32	(0.5) (3.3)	—	0.04–0.14	2/1
	403	25–33	32	(0.5) (3.3)	500–635	0.2–0.3	2/1
	602	66	32	(0.5) (3.3)	1000	1.2	4/4
	MPC860	25–40	32	(0.5) (3.3)	1800	0.54	4/4
Hitachi SH	7604	20	16–32	(0.8) (3.3)	450	0.24	4 unified
	7708	60–100	16–32	(0.5) (3.3)	800	0.4	8 unified
NEC	V810	25	32	(0.8) (3.3)	240	0.1	1 unified

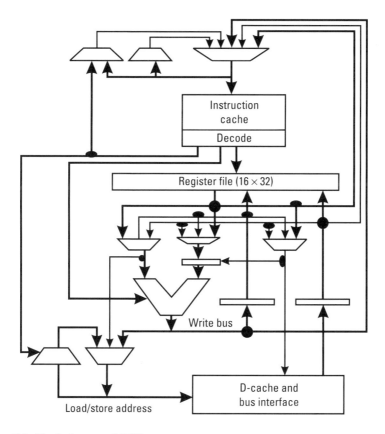

Figure 5.1 Block diagram of ARM core.

ASIC companies generally provide multiple versions of microprocessor cores in their library. As an example, five versions of the PowerPC 401 core are available from IBM for 3.3V, 0.35-μm technology:

1. 401B2: I-cache 16K, D-cache 8K, power dissipation 260 mW;
2. 401C2: 0 I-cache, D-cache 8K, power dissipation 190 mW;
3. 401D2: I-cache 4K, D-cache 2K, power dissipation 180 mW;
4. 401E2: no I-cache or D-cache, power dissipation 160 mW;
5. 401F2: I-cache 2K, D-cache 2K, power dissipation 170 mW.

Besides the microprocessor cores that have been developed from scratch, a growing trend is to convert older generation microprocessors into

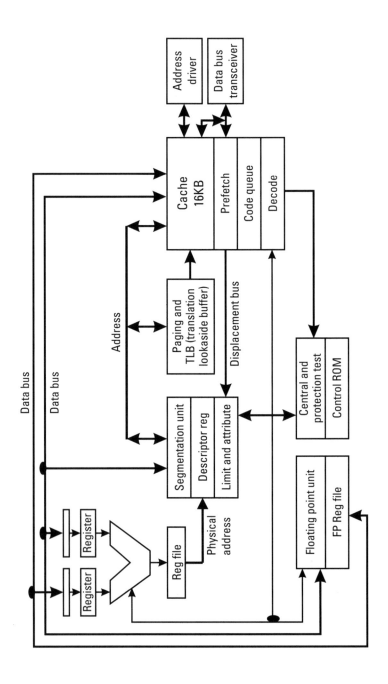

Figure 5.2 Block diagram of AMD's 486DX core.

reusable cores. This task seems simple; however, in reality it is quite a laborious and painstaking job. The architecture of a microprocessor core and the issues of converting an existing microprocessor into a reusable core are discussed in separate subsections.

5.1.1 V830R/AV Superscaler RISC Core

The NEC V830R/AV is a 32-bit superscaler and 64-bit SIMD superscaler RISC core [1]. Its internal clock speed is 200 MHz and off-chip communication is at 50 MHz. It contains a 16K four-way set associative I-cache and D-cache. The interfaces include Rambus DRAM interface through RDRAM control unit (RCU), video/audio interface, DMA, a four-channel 32-bit A-to-D multiplexed bus, and ICE interface for debugging.

The core architecture is shown in Figure 5.3 [1]; it is based on the V800 multimedia processor. It includes a V830-derivative integer execution unit with 32-bit MAC function, a 64-bit multimedia coprocessor, and a decoupled instruction unit. The CPU has a six-stage pipeline structure divided into three pipelines: instruction pipeline (I-pipeline), integer pipeline (V-pipeline), and multimedia pipeline (M-pipeline).

The M-pipeline has a 64-bit data path that acts as a multimedia coprocessor. Its arithmetic functions contain vector and scalar operations. It

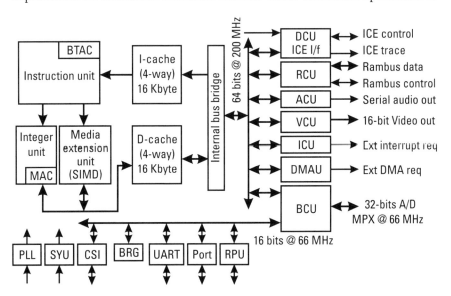

Figure 5.3 Architecture of NEC V830R/AV processor. (From [1], © IEEE 1998. Reproduced with permission.)

supports three types of multiply operations including one that is specifically targeted for the inverse discrete cosine transform (IDCT) required in MPEG-2 operations. It has a 64-bit execution data path with a partially reconfigurable SIMD execution capability. It is connected to a 32-word (64-bit) register file via 4-read and 2-write ports. The 1-read and 1-write ports are shared to transfer the load and store data from/to the D-cache; these ports are also shared with the V-pipeline.

The V-pipeline has a 32-bit data path that performs conventional functions. It has a 32-bit multiply-adder, 32-bit ALU, and a 64-bit shifter. It is connected to a 32-word (each 32 bits long) register file via 3-read ports and 1-write port; hence, it can execute 3-operand multiply-add instructions.

The design contains 3.9 million transistors. In the 0.25-μm, four-metal, 2.5V (3.3V I/O) CMOS process, it dissipates less than 2W.

5.1.2　Design of PowerPC 603e G2 Core

Reference [2] describes the conversion of the PowerPC 603e microprocessor into a reusable hard core named G2 in a 0.35-μm, five-metal process. The size of the G2 core is 34 mm^2 and it contains a 16K I-cache and 16K D-cache. The block diagram of the PowerPC 603e architecture is illustrated in Figure 5.4. Some notable items from this work are:

1. To obtain the necessary data transfer rate of internal buses, each bidirectional pin was redesigned to allow either partitioning into separate input and output core pins or remaining in a bidirectional configuration.

2. New sets of floor plan, place, and route constraints were developed. The aspect ratio was changed to square with pad connections to only the top left and bottom center of the floor plan. This was done to facilitate connections when this core is placed in one corner of a SoC.

3. Porous routing channels were used in the two top metal layers (more than 300 routing channels in the M4 and M5 layers) to allow over-the-core routing and SoC-level power grid.

4. Clocking and power buses were redesigned to allow the possibility of single-sided connections to Vdd and Gnd. Many design enhancements were made to reduce both the dynamic and static power consumption of the core.

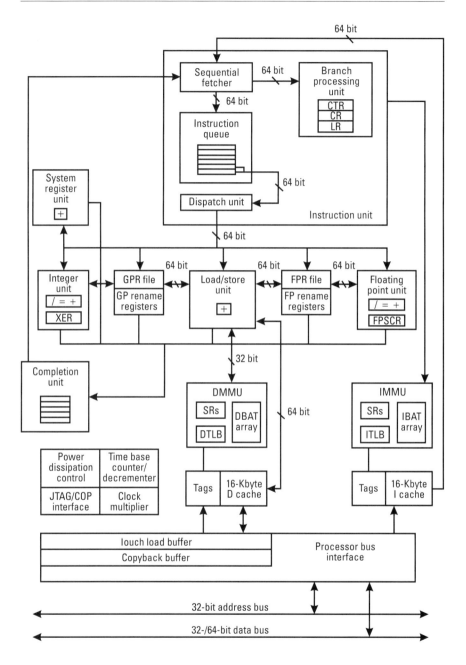

Figure 5.4 PowerPC 603e block diagram. (From [2], © IEEE 1998. Reproduced with permission.)

5. The simulation and design data formats were changed from proprietary formats to standard formats for reusability and SoC-level integration. The design was translated from proprietary language to Verilog, and a Boolean equivalence checker was used to validate the model. Additional integration rules were developed for reusability purpose. A testbench and procedure were developed to allow SoC designers to resimulate core patterns at the SoC level. Translation tables for the chip to core pin names were considered the responsibility of the SoC designer. The final design was delivered in GDSII format.

6. From a reusability point of view, test data were considered a key item. Hence, functional, scan, and embedded memory patterns were regenerated and verified. Embedded memories were tested by memory BIST controlled from the JTAG interface. Other enhancements for test and debugging included JTAG modifications so that JTAG ports can also work as debug ports and to allow multiple JTAG modules at the SoC level. A control was added to the core boundary to override chip-level interconnections so that a known functional mode can be used. This modification was also required in the boundary scan circuit so that the SoC designer can use a shell-type view of the core for chip-level ATPG run.

5.2 Comments on Memory Core Generators

As mentioned in Chapter 3, a majority of memory cores in SoC are developed through memory compilers. These memories include single and multi-port RAMs and register files optimized for either high density or speed. For example, memory compilers for various process technologies such as 0.18 and 0.25 μm from TSMC can be licensed from companies such as Artisan. To support these compilers, standard cell libraries and I/O libraries are also provided. Some example compilers are:

1. High-density single-port SRAM generator;
2. High-speed single-port SRAM generator;
3. High-speed dual port SRAM generator;
4. High-speed single-port register file generator;
5. High-speed two-port register file generator.

Each of these compilers provides a PostScript data sheet, ASCII data table, Verilog and VHDL model, Synopsys Design Compiler model; Prime Time, Motive, Star-DC, and Cadence's Central Delay Calculator models; LEF footprint; GDSII layout; and LVS netlist. A user can specify the number of words, word size, word partition size, frequency, drive strength, column multiplexer width, pipeline output, power structure ring width, metal layer for horizontal, and vertical ring layers.

5.3 Core Integration and On-Chip Bus

One of the major difficulties in SoC design is core-to-core communication infrastructure. Organizations such as the VSI Alliance have developed specifications for an on-chip bus interface to solve this problem. However, due to the lack of a standard and the various proprietary communication protocols associated with cores from different vendors, the plug-and-play concept is still far from reality. Attributes of some on-chip buses are discussed next.

IBM's processor local bus: This is a hierarchical bus architecture that contains a processor local bus (PLB), an on-chip peripheral bus (OPB), PLB arbiter, OPB arbiter, and PLB-to-OPB bridge. The PLB contains separate read and write data transfer handshaking, an address pipeline with decoupled addressing, and a data bus to support split-bus transactions. The data transfer protocol consists of request (initiated by master), transfer (arbiter grants bus), and address acknowledge (slave acknowledge and terminate address cycle). In a fully synchronous master/slave PLB controller arrangement, the master determines the length of burst transfer. The slave can terminate burst transfer and can force PLB rearbitration. The general structure is shown in Figure 5.5.

Hitachi SH bus interface: The Hitachi bus interface is tailored toward the Hitachi SH microprocessor core. It is also a hierarchical bus structure that requires a peripheral bus (P-bus), interface bus (I-bus), bridge module for connectivity of the H-bus to the peripheral bus, and bus personality interfaces (BPIs). The P-bus has a single bus master and allows single transfers of fixed duration with no pipeline. The H-bus allows multiple master operations, pipelined operations, and burst transfers of variable duration. It uses the PCI protocol and control signals for transfers. BPI is analogous to a wrapper around the core that is unique to each core. It may consist of address decode, data buffering, and clock synchronization functions. This structure is shown in Figure 5.6.

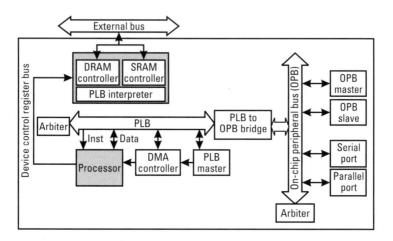

Figure 5.5 IBM's processor local bus-based architecture.

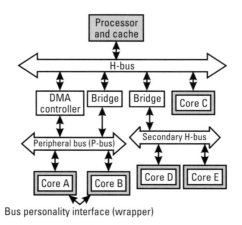

Figure 5.6 Hitachi's H-bus-based SoC architecture.

Toshiba general-purpose bus: This architecture contains an SoC-level general-purpose bus (G-bus), an X-bus internal to the R3900 processor, an interface module bus (IM-bus), and a G-bus-to-IM-bus bridge. The G-bus provides connectivity to the R3900 processor and to high-speed peripheral circuits such as a memory controller and interrupt controller, while the IM-bus is limited to low-speed peripheral circuits such as UART, PIOs, and timers.

The G-bus supports multiple-master, multiple-slave configurations, single word and burst modes, and two levels of arbitration (high and low priority). It also supports concurrency based on either snoop-and-transfer mode or execute-and-transfer mode.

Sonics integration architecture: The Sonics integration architecture consists of core interfaces (wrappers) that support, transmit, and receive FIFOs. These interfaces are connected to a backplane that supports arbitration, as either a computer bus or a communication bus. In computer bus mode, it uses a round-robin protocol with address-based selection and pipelining. In communication bus mode it uses the distributed TDMA communication protocol. Figure 5.7 illustrates the structure between transmitter and receiver.

Other examples of on-chip buses include ARM's advance microcontroller bus architecture (AMBA), ST Microelectronics PI bus, and the Inventra (Mentor Graphics) FISP bus. Proposals have also been made to use a PCI bus for core connectivity. Note that the Hitachi H-bus also uses the PCI protocol and control signals for data transfer.

5.4 Examples of SoC

In this section, two examples are given. The first example is of an SoC architectural design and the second is of SoC test methodology and testability. Although SoC test issues are discussed in Part II (from Chapters 6 to 10), we consider this a good place to present this example.

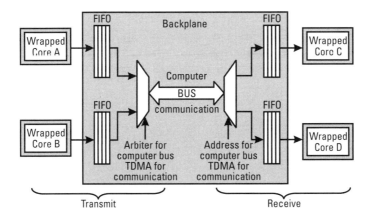

Figure 5.7 Sonic's backplane architecture for core connectivity and communication.

5.4.1 Media Processors

One good example of a present-day SoC is that of graphics accelerators and multimedia processors. An example of a multimedia processor (Philips Trimedia) is illustrated in Figure 5.8 [3]. Many of these processors perform MPEG-2 decoding for video. This decoding and playback of the compressed bit stream starts with variable-length decoding (VLD), followed by the inverse quantization (IQ) to retrieve discrete cosine transform (DCT) coefficients from the compressed bit stream, then an inverse DCT (IDCT) operation that produces the error prediction signal. This signal retrieves the decoded frame with the addition of a motion prediction signal. The motion prediction signal is calculated by pixel interpolation using one or two previously decoded frames. The decoded frame is transformed into a display format and transferred to the video RAM [3].

This architecture requires the five following functions:

1. *Bit manipulation function:* This is needed to parse and select bit strings in serial bit streams for variable length encoding and decoding.

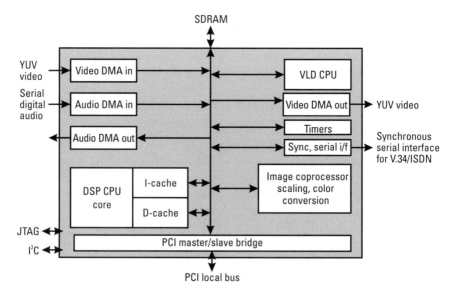

Figure 5.8 General architecture of a Trimedia processor. (From [3], © IEEE 1998. Reproduced with permission.)

2. *Arithmetic operations:* Subraction, addition, multiplication, and special arithmetic functions are needed for hardware performance and efficiency.

3. *Memory access:* Access to large memory space is needed to store the video frame. This also requires high bandwidth memory interface.

4. *Stream data I/O:* This is needed for compressed bit stream, video, and audio. For example, video requires capture and display as well as format conversion abilities.

5. *Real-time task switching:* This requirement imposes sample-by-sample or frame-by-frame time constraints such as in simultaneous video and audio decoding.

Today's CISC and superscaler microprocessors do not perform these tasks efficiently. For example, image processing has inherent parallelism in pixels and macroblocks. Although, superscaler control has a large amount of hardware, the issuing of two to four parallel instructions is not adequate to take advantage of the parallelism in image processing. Secondly, a microprocessor's word length has increased to 32 or 64 bits, whereas the word length required for multimedia processing is still at 8, 16, or 24. This difference causes large margins if multimedia data use arithmetic instructions on today's microprocessors.

From the memory-space point of view, video frames require a large amount of data—4.15 Mbits/frame to 0.5184 Mbits/frame for MPEG-2—that is too large for today's first-or second-level cache. Both audio and video require that the processing be completed within a predetermined sample or frame interval time. Data-dependent operations and memory accesses prohibit using the cache to predict the processing cycle in microprocessors.

To overcome the microprocessor's limitations, multimedia processors have been developed. Generally, these multimedia processors use the following components:

1. *Microprocessor core:* Generally, an 8- or 16-bit embedded microprocessor core is used. For example, an RISC core that matches the word length requirement in multimedia processing.

2. *DSP core:* A DSP core is chosen for bit manipulation functions as well as to enhance the performance and efficiency of arithmetic operations.

3. *Large memory:* A large DRAM block such as 16 Mbits allows the storage of both receive and transmit frames. On-chip storage also provides higher bandwidth and reduced delay.

4. *Video and audio DACs:* Both audio and video DACs are required to output the processed bit stream. Video DACs are triplet DACs for the RGB to YUV range.

5. *Interface controller cores:* These are required for high-speed memory interfaces such as SLDRAM and Rambus, as well as standard interfaces such as PCI, USB, I²C, and UART.

6. *Special-purpose circuitry:* This circuitry may include application-specific cores such as a fax/modem core, a temperature monitor, and also test/debug and access circuitry such as JTAG (IEEE 1149.1) and ICE.

As an example of this architecture, a block diagram of a Sony/Toshiba media processor is given in Figure 5.9(a) [4]. This media processor contains a superscaler RISC core based on the MIPS-III instruction set, which includes most of the MIPS-IV ISA and 107 additional integer SIMD instructions.

This core has thirty-two 128-bit registers that are used by the two 64-bit integer units. The core can issue two instructions for each cycle. Branch prediction is done via a 64-entry branch target address cache. The core has a 16K two-way associative I-cache, an 8K two-way set associative D-cache, and a 16K scratchpad RAM (SPR). Two 48-entry translation look-aside buffers (TLB) translate virtual addresses to physical memory addresses. SPR is logically in a separate memory space, enabled by an SPR-flag bit in the TLB entries. The memory architecture is double-buffered local memory with a DMA controller.

The Sony/Toshiba processor contains two vector processor units (VPU0 and VPU1), which are based on the VLIW-type SIMD architecture with floating-point units. The block diagram of VPU is given in Figure 5.9(b) [4]. While VPU0 functions as a coprocessor to the RISC core to perform behavior and physical simulation, VPU1 works independently for fixed, simple geometry operations that generate display lists to be sent to the rendering engine. To avoid disturbing other operations on the main bus during data transfer, VPU1 has an exclusive path to the rendering engine. VPU0 on the other hand stores its results temporarily in SPR before they are transferred to either the rendering engine or VPU1. Each VPU has four FMAC and one divider and a subprocessor elementary function unit that has an additional FMAC and divider (for both VPU, there are 10 FMACs and four

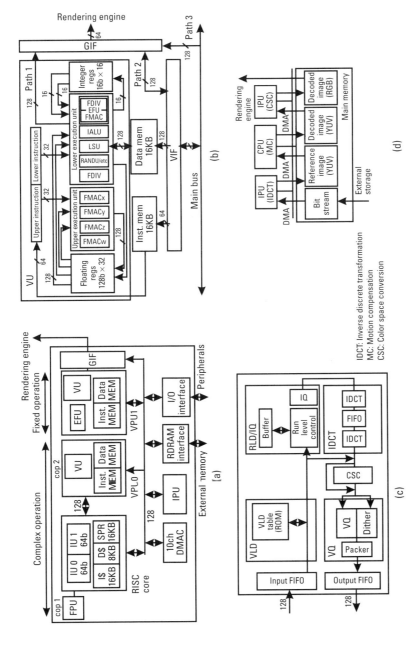

Figure 5.9 Sony/Toshiba media processor: (a) architecture; (b) VPU block diagram; (c) IPU block diagram; and (d) MPEG-2 block diagram of video decoders. (From [4], © IEEE 1999. Reproduced with permission.)

dividers). Image processing is done by MPEG-2; however, IPU is not exclusive for MPEG-2 video decodes. Block diagrams of IPU and MPEG-2 video decoders are illustrated in Figures 5.9(c) and (d), respectively [4].

The chip peripherals include a 10-channel direct memory access controller (DMAC), image processing unit (IPU), memory interface, I/O interface, graphics interface unit (GIF), and two RDRAM channels. These units are connected through an on-chip shared 128-bit bus to the RDRAM memory controller.

As an example of physical design, the Sony/Toshiba processor's floor plan with a clock tree—which is based on a three-level buffered H-tree—is illustrated in Figure 5.10 [4]. The clock is distributed through the root buffer, three levels of global buffers, and two levels of local buffers. The target clock skew was 120 ps (40-ps global, 60-ps local, and 20-ps margin). The global clock wires are routed in the top metal layer that is limited to clock, power, and ground. The chip runs at 250-MHz clock speed with power dissipation of about 15W at 1.8V that is estimated as 42% in CPU core, 44% in vector and other units, and 14% in I/Os. The metal width of the internal

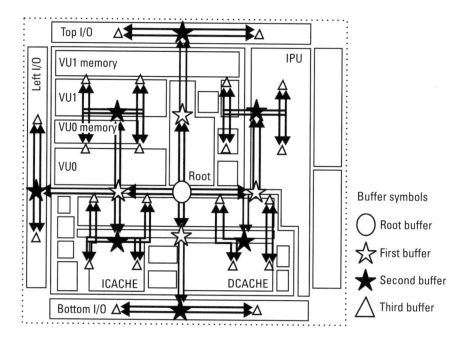

Figure 5.10 Clock tree in Sony/Toshiba media processor. (From [4], © IEEE 1999. Reproduced with permission.)

power lines was defined considering voltage drop and electromigration rules. Overall the chip contains 10.5 million transistors fabricated in a 0.25-μm (0.18-μm drawn), 1.8V process. The 17×14.1-mm^2 die is housed in a 540-pin PBGA package.

5.4.2 Testability of Set-Top Box SoC

In 1998, Fujitsu conducted a socket program to demonstrate its SoC test methodology. Reference [5] describes the testability of this pilot project, which was based on the MPEG2/JSAT3 chip (MB86372) for validating the concepts of a BIST friendly test methodology. The following testability was used for various cores/blocks and user-defined logic:

1. *Viterbi decoder core:* Consists of about 100K gates and 10 memories. For memories, the memory BIST controller was used inside the core. Original RAMs were replaced by RAMs with wrappers. Logic was analyzed for scan test points, and scan chains were inserted. A wrapper was used for the whole core and internal scan chains were controlled by logic BIST controller.

2. *GPIO core:* This is a legacy core for which functional patterns were available. A wrapper was generated and integrated; and functional vectors were verified through this wrapper.

3. *MIU block:* This consists of about 40K gates and 10 memories. Again, a memory BIST controller was used inside the block to test memories. Original memories were replaced by memories with wrappers. Scan chains and test points were inserted to test the logic.

4. *TSD block:* This has about 20K gates and 3 memories. Wrappers were generated and integrated with memories, while a chip-level memory BIST controller was used. For logic, again, test points and scan chains were inserted.

5. *AO block:* This consists of about 20K gates and 3 memories. The memories were wrapped and a chip-level memory BIST controller was used. For logic, test points and scan chains were inserted.

6. *VO block:* This consists of about 20K gates and 3 memories. Memories were wrapped and a chip-level memory BIST controller was used. For logic, test points and scan chains were inserted.

7. *ICC block:* This has about 10K gates of logic for which test points and scan chains were inserted.

8. At the chip level, a memory BIST controller for TSD, AO, and VO blocks was integrated. A logic BIST controller, boundary scan, and TAP controller were also integrated.

The test area overhead was about 22% of the chip size. However, it provided a chip-wide test-access mechanism and test automation at the core level as well as SoC level.

At this stage, we suggest that the reader refrain from being biased toward any core/SoC test methodology based on this example. In Part II, Chapters 6 to 10, we will discuss various test methodologies for microprocessor cores, application-specific cores, memory cores, and analog/mixed-signal cores. We also discuss the manufacturing test aspects in Chapter 10.

References

[1] Suzuki, K., et al., "V830R/AV: Embedded multimedia superscaler RISC processor," *IEEE Micro,* March–April 1998, pp. 36–47.

[2] Hunter, C., and J. Gaither, "Design and implementation of the G2 PowerPC™ 603e embedded microprocessor core," *Proc. IEEE Int. Test Conf.,* 1998, pp. 473–479.

[3] Kuroda, I., and T. Nishitani, "Multimedia processors," *Proceedings of the IEEE,* Vol. 86, No. 6, June 1998, pp. 1203–1221.

[4] Suzuoki, M., et al., "A microprocessor with 128-bit CPU, ten floating point MACs, four floating point dividers and an MPEG-2 decoder," *IEEE J. Solid State Circuits,* Vol. 34, No. 11, Nov. 1999, pp. 1608–1616.

[5] Zorian, Y., D. Burek, and S. Mukherji, "System-on-chip test: methodology and experience," IEEE P1500 web site, http://grouper.ieee.org/groups/1500/.

Part II:
Test

6

Testing of Digital Logic Cores

While the use of cores in system chips serves a broad range of applications, these chips are far too complex to be tested by traditional methods. In the production of SoC, a combination of test methodologies is used such as a functional test, full-scan test, logic BIST, RAM BIST, and Iddq. In a broader sense, the individual cores are tested by one or more of the following methods [1, 2]:

1. Testing a core through the system chip's functional test;
2. Direct test application while accessing the core through I/O muxing;
3. Test application to core through a local boundary scan or collar register;
4. Full-scan and built-in self-test through a variety of access and control mechanisms;
5. Proprietary solutions.

In this part, we discuss these methods in detail. But first, we need to discuss the issues in testing that are unique and that make SoC testing far more complex than traditional ASIC testing.

6.1 SoC Test Issues

Numerous difficulties surround the implementation of the methods just listed for testing cores and the testing of SoC as a whole. In [3], the following SoC testing problems were identified (listed in arbitrary order):

- Timing reverification;
- Lack of scan and BIST;
- Getting at embedded pins;
- Combining differing test methodologies from different IP providers;
- At-speed testing;
- Test pattern reliability for hard cores;
- Access;
- Controllability;
- Observability;
- Reuse of test;
- Reuse of DFT;
- Mixed signal test;
- Lack of Boundary scan;
- Synchronization.

A wide range of circuits come within the definition of digital cores. These circuits may be microprocessors/microcontrollers, DSP cores, and function-specific cores such as a modem or Internet tuner, as well as bus controller and interface circuits such as PCI, USB, and UART. In the following section, we discuss the general architecture and test methods used for digital logic cores.

If we start from the definitions of soft, firm, and hard cores, it is clear that the definition as given in Section 1.1 emphasizes the reusability of the design. The reuse of cores provides high design productivity; it automatically requires equivalently high test productivity and thus necessitates *test reusability*, which is a difficult problem to solve. Test reusability implies the development of a test strategy and incorporation of structured design-for-test (DFT) features and the generation of test vectors in a universally acceptable form. All chip integrators have in-house product development methodologies that consist of various EDA and in-house CAD tools. Test reusability implies that the core provider should consider (and possibly support) all of these tools and methodologies.

The difficulty is further compounded because the core provider has no way of knowing all possible situations in which the core might be used. This means that there should be very little to no constraint associated with the *core test strategy* so that the same core test can be used in any situation without much difficulty. It is thus linked to another issue: *integration of the core test* with the chip test. Note also that because the core provider cannot anticipate all possible core usage situations, the controllability and observability of the core is also unknown. Hence, the core provider needs to develop a core test strategy that imposes minimal access, control, and observation constraints on the chip integrator.

Core test integration is as complex a problem as test reusability for the same reason: The core provider has no way of knowing the end application, tools, and methodologies used by the chip integrator. Let us consider the example of a soft core when the core design is not synthesized. In this case, the chip integrator can easily modify the core test and even add some (DFT and BIST) features if necessary. For all practical purposes, the core test becomes the responsibility of the chip integrator; only simple test guidelines are sufficient and expected from the core provider. However, in the case of a hard core, in which the whole design (all the way to layout) is fixed and modifications are not possible, the test methodology used by the core provider must be developed such that it will not break down at the chip level. In general, the core provider provides a test set with the hard core, and the core integrator needs to integrate it at the chip-level test, often without fully understanding the operation/execution of the core test. Hence, issues such as *isolation* during the core test, *access mechanisms, test control, and observation mechanisms* come into the picture.

In the context of embedded core testing, *isolation* involves electrically detaching the input and output ports of the core from the chip logic (or other core) connected to these ports. Isolation may be required in two cases: during internal testing of the core and during testing of the logic outside of the core (other cores or user-defined logic, UDL). During the testing of the core, it may be necessary to ensure that the UDL or other cores do not interfere with the test of the core. During the testing of UDL, it may be necessary to ensure that the core logic and core inputs are protected from the effects of these tests. The isolation can be input isolation, output isolation, or both.

6.2   Access, Control, and Isolation

The general discussion just given identifies some key test difficulties in the testing of cores; all of these difficulties are applicable to logic cores. The fundamental items in SoC testing are *access, control,* and *isolation.* The test access mechanism refers to the application of the desired test stimulus at the input of the core and getting the response from the core under test. To some extent this is equivalent to (probably a subset of) the traditional terms *observability* and *controllability.* The term *control* in core testing refers to the mode that activates or deactivates the core testing functions. For example, if a core contains built-in self-test (BIST), the mechanism used to activate BIST and switch back to normal functional mode is referred to as the control mechanism. The chip integrator has to make proper provision for access to core test control mechanisms and also provide methods to activate and deactivate them.

The new term that is not used in traditional VLSI testing is *isolation,* which refers to electrically detaching the input and output ports of the core from the chip logic (or other cores) connected to these ports in order to avoid any adverse effects on the UDL or other cores. Isolation may be required during core testing as well as during UDL testing.

Some desirable features of core isolation and the core-test access mechanism at the SoC level are as follows:

1. Allows delivery of test data to and from the core without adverse interference from other cores or UDL.

2. Prevents the process of delivery of test data from interfering or damaging surrounding cores or UDL.

3. Allows simultaneous testing of several cores in isolation from each other.

4. Allows verification of interconnect wires between cores and between the core and the UDL.

5. It should be based on a simple methodology to ensure that the core can be tested without implying any requirements on neighboring cores or UDL.

The main objective of isolation is to avoid any adverse effects on neighboring circuits (other cores or UDL) as well as to protect the core when neighboring circuits are being tested. Depending on the SoC structure, isolation might be applied only to the inputs of the core, only to the outputs of the core, or to both. When two cores share some signals, and isolation logic is used at only one core, it is referred to as *single isolation*; when isolation logic is used at both cores, it is referred to as *double isolation.*

When a multiplexing structure is used to obtain isolation, the same isolation logic also provides test access functions. Figure 6.1(a) illustrates a single isolation mechanism using one mux at the input side. In this structure, one can force test data at the input of the core and read the results after UDL. The only item to be cautious about here is that the core test data propagate into UDL and, thus, testing of multiple cores in parallel may be difficult or impossible. Another possibility is to reverse the point of stimulus and observation, as shown in Figure 6.1(b). In this case, the observation points are at the output of the core and the test data need to propagate through UDL before arriving at the core. The limitation in both Figures 6.1(a) and (b) is that interconnect tests can be very difficult unless more observation points are made. Note also that in Figures 6.1(a) and (b) all signals are registered to obtain isolation, but in reality there are always some exceptional signals such as clocks that are not registered.

The limitation of Figures 6.1(a) and (b) can be overcome by double isolation as shown in Figure 6.1(c). In this case, observation and control occur at both outputs and inputs. The control point substitutes the output during testing and the observation point carries output data to a transport mechanism or to the primary output. With this method, the core provider can ensure that the core can be tested in a stand-alone manner with no dependencies on its surroundings. However, because of the multiplexers at both inputs and outputs, it results in a net performance penalty as well as higher area overhead.

6.3 IEEE P1500 Effort

To address the wide range of test issues such as test access, test control, and the observation mechanism for embedded cores, the IEEE Technical Council on Test Technology has established a working group (P1500) to develop a standard architecture that will solve these issues. The basic principles of P1500 scalable architecture follow [4].

The embedded core test, in general, requires the following hardware components: (1) a wrapper (around the core), (2) a source/sink for test patterns (on- or off-chip), and (3) an on-chip test access mechanism to connect the wrapper to the source/sink.

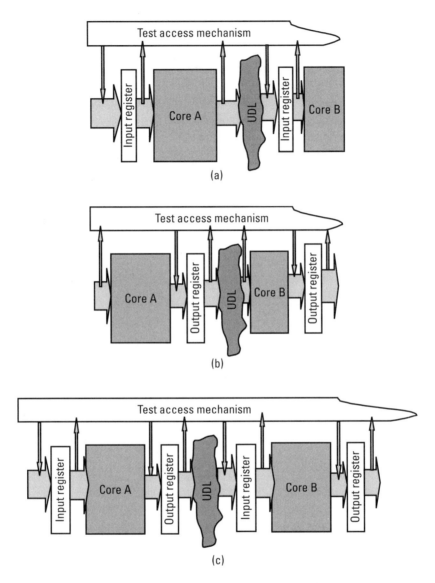

Figure 6.1 Illustration of isolation methods: (a) input isolation; (b) output isolation; and (c) double isolation.

The P1500 Scalable Architecture Task Force is defining the behavior of a standard wrapper to be used with every core that complies with P1500 and an interface between the wrapper and the test access mechanism. The objective of this wrapper is to facilitate core test, interconnect test, and isolation functions by providing switching between different test and diagnostic modes and the normal functional mode.

The wrapper of a core may interface with three types of input/output signals:

1. Static control signals for wrapper modes;

2. Digital data and dynamic control signals (to be routed through wrapper cells);

3. Exceptional signals (to bypass wrapper cells), such as clocks, high-speed, asynchronous, analog, and other special-purpose signals.

Standard wrapper behavior can be implemented and provided by the core vendors or it can be added to the core at one of the design stages of SoC. The assumption is that EDA vendors will offer tools to implement the standard wrappers, check for compliance, and provide system chip-level optimization.

The interface between the wrapper and the test access mechanism will also be standardized, but a specific test access mechanism will not be defined for SoC because this implementation depends on the test strategy used by the SoC designer. One compliant implementation of the P1500 standard interface and wrapper might be the 1149.1 TAP interface and its boundary-scan register.

To achieve the above principles, P1500 will possibly consider the following requirements.

With the design, simulation, and test vector files of intended functionality of the core, the core provider must do the following: augment the functionality of the wrapper into the core and provide an RT-level simulation model of its behavior, or provide a file of an RTL or gate-level design with synthesis script and a text description of its operation, which if implemented with the core design according to the text description will perform the P1500 functions. The P1500 functions ensure the following:

- An isolation such that during the core-test mode any activity in the core and wrapper will not adversely affect any other section of the chip;

- Access to the core design to use the testability features, test vector application, and test response observation;

- Access to interconnects and UDL between cores or between a core and primary I/Os, such that any testing of interconnects and UDL will not adversely affect the core circuitry.

The core provider must also provide a file compatible with P1500 Core Test Language (CTL) that describes the core's testability features, operation and control of these features, the method to apply the test vectors, and the exact sequence of operation of the core testing.

Note that the requirements just discussed may or may not be mandatory for the core provider; thus, the core integrator may or may not use these files for optimization purposes or otherwise. Also, these requirements do not specify an implementation method for the wrapper.

At this time, it is expected that the core-test standard information model (test patterns) will be defined by the P1500 CTL group, and the standard wrapper and interface to the test access mechanism will be defined by the P1500 Core Test Scalable Architecture (CTAG) group [5–7]. The P1500 standard is also expected to be applicable to hierarchical cores.

6.3.1   Cores Without Boundary Scan

With structures like those shown in Figure 6.1, a generalized architecture can be developed for testing cores as shown in Figure 6.2 [7]. This architecture is under discussion by the IEEE P1500 working group as a possible standard for the core providers. In Figure 6.2, a core is shown with full scan (two parallel scan chains), although the architecture permits any DFT/BIST implementation or no DFT/BIST and only functional vectors. The core also has boundary-scan-type collar registers at the inputs and outputs, which are accessible through serial data-in (SI), serial data-out (SO), or parallel data-in (PI) and parallel data-out (PO). These registers select functional data-in ($d[0:4]$), functional data-out ($q[0:2]$), or test data through SI/PI and SO/PO. Note that Figure 6.2 contains one possible example of cells for collar registers; many other implementations are possible.

The reasoning behind serial and parallel test mechanisms is that for the same core some SoC designers might prefer serial access, whereas others might want parallel access. It allows the SoC designer to decide on the trade-off among test time, area overhead, routing, and performance penalty. For example, serial access requires a single wire with a serial control interface and mode select signal, while parallel access can have an n-bits-wide bus with parallel control interface. A bypass register (similar to the bypass register

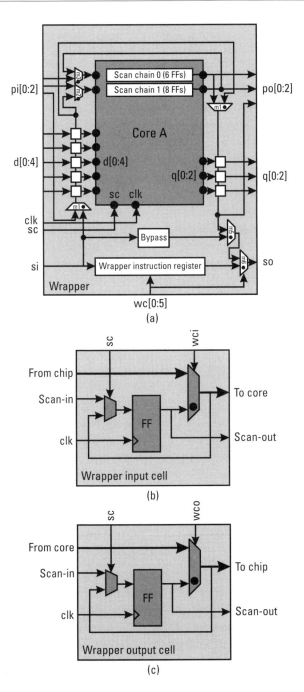

Figure 6.2 (a) A generalized architecture for core testing. (b) Example of input cell. (c) Example of output cell. (From [7], © IEEE 1999. Reproduced with permission.)

specified in the IEEE 1149.1 standard) is also implemented to allow transportation of data through a core without disturbing its logic.

Note that the SoC-level integration of such a core requires an SoC-level scan path to access SI/PI and SO/PO. Because boundary-scan-type operations are needed, this controlled scan path (CSP) needs logic that can perform IEEE 1149.1 type shift/update/capture operations. Considering these requirements, one possible implementation of CSP is given in Figure 6.3(a) [4]. A possible topology of CSP usage at SoC level is illustrated in Figure 6.3(b) [4]. Note that the example of CSP implementation shown in Figure 6.3(a) not only provides 1149.1-type operations for logic testing, it also facilitates Iddq testing of the core. Further details on Iddq testing of the core and SoC are given in Chapter 9.

It is also worth mentioning that the additional logic proposed in this architecture will result in a net performance penalty. Thus, discussion is taking place about whether the core provider should provide separate files for the functional logic of the core and additional logic (BSCAN-like collar

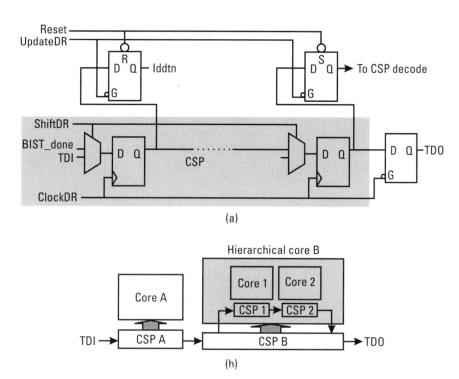

Figure 6.3 IEEE P1500 control scan path under discussion: (a) one possible implementation and (b) a possible use at the SoC level. Based on [4].

registers, test points for serial/parallel access, bypass register, and so on), so that SoC designers can optimize test logic at the chip level if necessary.

6.3.2 Core Test Language

To facilitate core test reuse and SoC-level test program development, the P1500 working group has also established a Core Test Language (CTL) Task Force. This task force is evaluating STIL [8] and making some enhancements to satisfy the core test needs [4]. The preliminary evaluation has suggested that with some enhancements STIL may be adequate to describe various intricate signals, hierarchical models, and complex signal timings required by the core test and its reuse at the SoC level. The STIL description of the arbitrary example given in Figure 6.2 is illustrated in Figure 6.4. In this example, a module with two scan chains is used; the module also has a variety of control and data signals. Using some assumed values, the description in Figure 6.4 shows that with some enhancements, the STIL language is adequate to describe core test. It allows for the passing of parametric values to the macro and it can also express vector data, inversion in the vector data, multiple shift statements within a macro, and complex timing.

6.3.3 Cores With Boundary Scan

The architecture discussed in Section 6.3.1 contains a number of elements and operational behaviors that are similar to the 1149.1 boundary scan standard [9]. To ease the SoC-level integration effort, companies such as Texas Instruments have developed core design methodologies that are in agreement with IEEE 1149.1 standards. These cores contain a localized TAP controller and a boundary-scan register that behaves as a collar to the core. The IEEE P1500 working group is also targeting both non-TAP and cores with localized TAP controllers, and the P1500 proposal is expected to support both types of cores. An example of various operations through the wrapper is shown in Figure 6.5 [7].

References [10] and [11] discuss the SoC design styles based on cores with a localized boundary scan and TAP controller. With every core containing a localized boundary scan and TAP controller, the integration of cores at the SoC level as well as testing is quite simplified. In this situation, a master TAP controller at the chip level controls the operations of individual TAP controllers of the cores either through a TAP linking module or a switching circuit. Thus, the SoC resembles a board, and each core resembles an IC on

```
CTL 1.0                                         CoreInternal { DriveAccuracy NonCritical; }
Header {                                         CoreExternal {
 Title "CTL Program for 1500-compliant Core";    ConnectTo PIN;
 Date "Fri June 25 1999";                         DataType TestControl;
}                                                 }
CoreSignals core_a_bare_signals {               }
 // original core signal definition is maintained }
 d[0..4] In { CoreExternal { ConnectTo UDL; }}  SignalGroups {
 q[0..2] Out { CoreExternal { ConnectTo UDL; }}  si_m123 = 'si' { ScanIn 22; }
 si0 In { ScanIn 6; CoreExternal { ConnectTo TAM; }}        // length of (d+ch0+ch1+q)
 si1 In { ScanIn 8; CoreExternal { ConnectTo TAM;}}  si_wir = 'si' { ScanIn 6 }
 so0 Out { ScanOut 6; CoreExternal { ConnectTo TAM; }}  so_m123 = 'so' { ScanOut 22; }
 so1 Out { ScanOut 8; CoreExternal { ConnectTo TAM; }}  wc = 'wc[0..5]';
 sc In { CoreExternal { ConnectTo PIN; }}       }
 clk In { CoreExternal { ConnectTo PIN; }}      Timing normal {
}                                                // waveform and timing details omitted
Signals {                                        WaveformTable setup_timing { }
 // New Signals block for the wrapped core       WaveformTable do_intest_timing { }
 si In {                                        }
  CoreExternal { ConnectTo TAM; }               MacroDefs {
 }                                               setup_interest {
 sc In {                                          Purpose Instruction;
  CoreExternal { ConnectTo PIN; }                W setup_timing;
 }                                                // setup default conditions on all signals used
 clk In {                                         C { si=0; so=X; wc=000000; }
  CoreInternal { DriveAccuracy Critical; }        // put WIR in receive mode
  CoreExternal { ConnectTo PIN; DataType Clock; }  V { wc[0..4]=10000; }
 }                                                // shift instruction into WIR
 d[0..4] In {                                     Shift { V(si_wir=111000; wc[5]=1;} }
  CoreInternal {                                  // put WIR in update mode; instruction in effect
   ConnectIn Scan mlm4 '0..4';                    F { wc=010001; }
   Wrapper IEEE-1500;                            }
  }                                              do_intest {
  CoreExternal { ConnectTo UDL; }                 Purpose ControlObserve;
 }                                                W do_intest_timing;
 pi[0..2] In {                                    C { si_m123=0, so_m123=X; clk=1; sc=1;
  CoreExternal { ConnectTo TAM; }                 Shift { V(si_m123='d, si1, si0';
 }                                                        so_m123='so1, so0, q';} }
 po[0..2] Out {                                   V { clk=1; sc=0 }
  CoreExternal { ConnectTo TAM; }                }
 }                                              }
 q[0..2] Out {                                  PatternExec run_all {
  CoreInternal {                                 Purpose production;
   ConnectOut Scan mlm4 '5..7';                  Timing normal;
   Wrapper IEEE-1500;                            Category normal;
  }                                              PatternBurst all_pats;
  CoreExternal { ConnectTo UDL; }               }
 }                                              PatternBurst all_pats {
 so Out {                                        TestMode InTest;
  CoreExternal { ConnectTo TAM; }                PatternSet { pat1; pat2; pat3; }
 }                                              }
 wc[0..5] In {                                  Include "pat1.stil";
```

Figure 6.4 Example of core's test development in P1500 CTL. (From [7], © IEEE 1999. Reproduced with permission.)

the board as shown in Figure 6.6 [11]. If the chip also has non-TAP cores (NTC), this master TAP controller can facilitate communication to them.

The state diagram of an individual core's TAP controller is exactly like that described in the IEEE 1149.1 standard; however, the state diagram of the master TAP controller requires control states followed by update-DR and update-IR as shown in Figure 6.7(a) [11]. As an example, Figure 6.7(b) shows a timing and state diagram for control transfer from the master TAP to the core's TAP [11].

The advantage of this approach is that it provides a very well-defined structure and well-understood communication protocol, and it utilizes prior

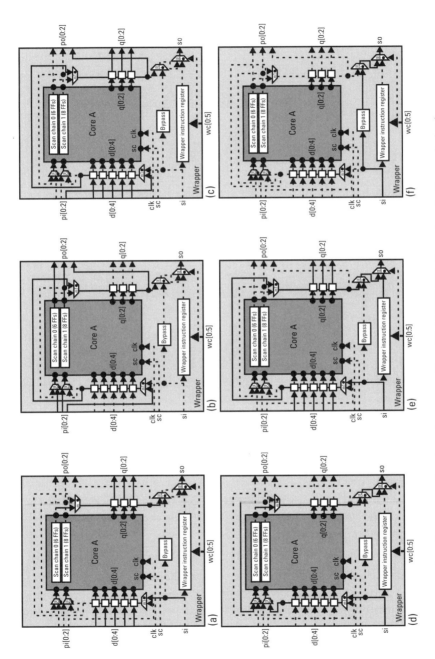

Figure 6.5 Different modes of core operation through wrapper. (From [7], © IEEE 1999. Reproduced with permission.)

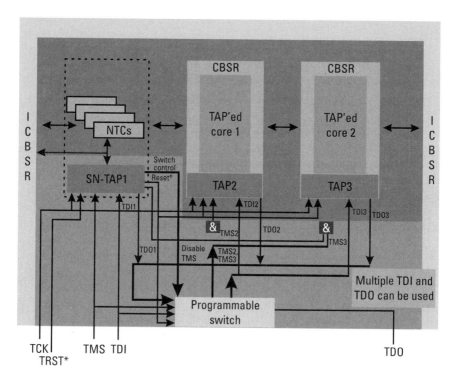

Figure 6.6 SoC design using cores with localized TAP controllers. (From [11], © IEEE 1998. Reproduced with permission.)

knowledge of boundary scan. The main limitation to this approach is the fact that the testing of every core as well as the chip-level interconnect is done through boundary scan. The test time is thus significantly large and at-speed testing is not possible.

6.4   Core Test and IP Protection

In core-based SoC, one significant concern for the core providers is the protection of intellectual property (IP), particularly for the firm and hard cores. The concern ranges from piracy and illegal use to reverse engineering. To address piracy and illegal use of cores, the VSI Alliance plans to develop an IP protection specification using various techniques such as a digital signature, watermarking, and use of antifuse technology for both ASICs and FPGAs [12–14].

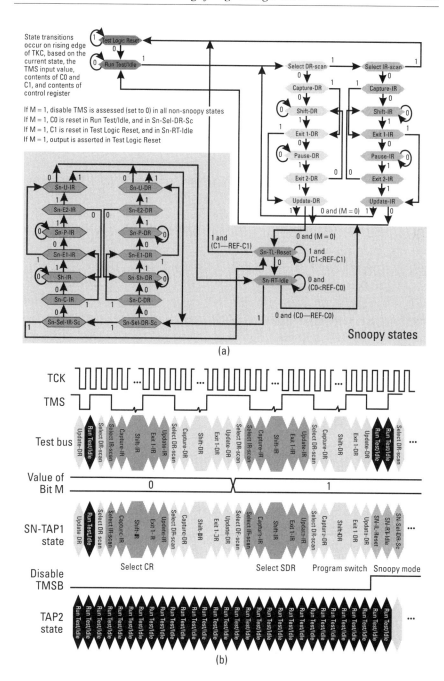

Figure 6.7 Operation of hierarchical TAP controller: (a) state diagram of master controller and (b) example of timing diagram when master TAP yields test bus to core's TAP. (From [11], © IEEE 1998. Reproduced with permission.)

The main issue for core providers in releasing the core netlist is the potential loss of a proprietary design. The value of IP is generally in terms of how the design has been implemented because implementation defines the area and performance of the core. In the case of firm cores, the core provider is asked to provide a gate-level netlist of the core; in the case of hard cores both a gate-level netlist and layout database are required so that the core can be integrated at the SoC level.

The gate-level netlist contains all of the structural information necessary to understand the performance and area of the core. In other words, almost all of the IP value related to design implementation is available with the gate-level netlist of the core. The core providers have considered encryption methods to protect the gate-level netlist. However, encryption methods have not been successful so far because an unencrypted gate-level netlist is still required by the ATPG for test generation at the SoC level.

A proprietary method developed at LSI Logic solves this issue [15]. With this method, the core provider can delete some amount of gate logic from the gate-level netlist. The idea is to release only that part of the gate-level netlist that is absolutely required by the ATPG for test generation of the whole design. In other words, if a gate is outside the observability and controllability cones (cone from core's input to core's output or scan flip-flops), its presence and Boolean properties are not required during chip-level test generation and, hence, it can be deleted from the released gate-level netlist. Note that these deleted gates are not redundant gates; they are not deleted from the actual implementation of the core, they are only deleted from the gate-level view of the core that is required to be released. This modified gate-level netlist (partial netlist) is released for test generation at the chip level along with a small test shell necessary to test the gates, which are in the actual implementation but not in the released gate-level netlist.

The basic steps to identify gates that can be deleted are given below; Figure 6.8 shows an example to illustrate the steps.

1. Perform a backward traversal from the core's outputs and mark all the gates in the paths from the core's outputs to the core's inputs or scan flip-flops. These gates constitute the input cones of the core's output and are required for observability reasons at the core's output.

2. Perform a forward traversal from the core's inputs and mark all the gates in the paths from the core's inputs to the core's output or scan flip-flops. These gates constitute the output cones of the core's

input and are required for controllability reasons at the core's inputs.

3. Mark all of the gates that are required for logic propagation through any of the gates marked in step 1 or step 2 (input cones or output cones). These are referred to as side gates.

4. Any gates not marked in steps 1, 2, or 3 can be deleted because they are not necessary for the controllability of the core's inputs or observability of the core's outputs.

5. The number of marked gates in steps 1 to 3 can be reduced by connecting some of the core's inputs and outputs in a boundary-scan-type register (shift register to provide controllability/observability at the core's I/Os during testing). A heuristic algorithm is also given in [15] to select a small number of core I/Os for this purpose.

Figure 6.8(a) shows an example circuit, whereas Figure 6.8(b) shows the marked gates in steps 1, 2, and 3. In step 1, gates 10, 11, 12, 13, and 14

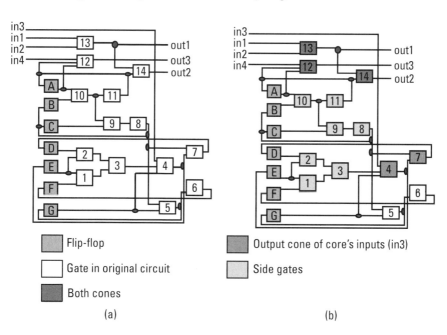

Figure 6.8 Example to illustrate gate-level netlist reduction method: (a) original circuit; (b) marked gates in steps 1, 2, and 3. Gates 5 and 6 in part (b) can be removed. (From [15], © IEEE 1997. Reproduced with permission.)

are marked; in step 2, gates 4, 7, 12, 13, and 14 are marked; in step 3, gates 1, 2, 3, 8, 9, 10, and 11 are marked. Gates 5 and 6 are thus identified as unnecessary for both observability and controllability. Therefore, these two gates can be deleted to obtain a partial netlist that is sufficient for chip-level test generation. In other words, if the partial netlist (without gates 5 and 6) is merged into a chip-level netlist, the ATPG will not have any difficulty propagating a fault effect on inputs 1 to 4 to outputs 1 to 3. In addition, this partial netlist will not cast a shadow for ATPG after outputs 1 to 3.

As observed in [15], for many cores, steps 1 to 3 identify only a small number of gates that can be deleted from the gate-level netlist. However, with about 20% of the core's I/Os in a boundary-scan-type connection to enhance observability/controllability at these I/Os, a 30% to 80% gate-level netlist can be successfully deleted from the majority of cores for release purposes. Enhancing core testability by using a partial isolation ring has also been studied by other researchers [16]. Besides the method discussed in [15] to select core I/Os for a partial boundary scan, a method such as that given in [17] can be used to gain access from the UDL.

6.5 Test Methodology for Design Reuse

For design and test reuse, a number of core providers and ASIC companies have established full-scan and boundary-scan-based test methodologies for their cores.

6.5.1 Guidelines for Core Testability

In most of the cases, core test methodologies used by ASIC manufacturers have the following characteristics:

- Fully synchronous design;
- More than 95% functional fault coverage;
- Full-scan and combinational ATPG for more than 95% stuck-at coverage;
- A small number of Iddq vectors (generally <100);
- Built-in self-test with shared controller for all memories (>4 Kbits) embedded in the core (For memories of less than 4 Kbits, used as a register file, muxing occurs through the core's I/O boundary scan

cells. All memories, BIST or otherwise, were tested by March algorithm of $9n$ or higher complexity.);

- Boundary-scan-type register at the core's I/Os; however, the netlist of the boundary scan is deliverable as a separate file from the core netlist, under top-level module netlist;

- Functionality debug flow;

- Test vector files and timing diagrams;

- Test suite documentation including process monitor information if the core contains an embedded process monitor cell;

- Core silicon bond-out for emulation and a stimulus file for emulation.

6.5.2 High-Level Test Synthesis

As discussed in Section 6.2, the core test methodology must facilitate test reuse. Test synthesis at the behavioral or RT level can be significantly helpful. For soft and firm cores in particular, RT-level test synthesis is ideal because it creates the smallest number of constraints for chip integrators and facilitates test reuse for different core applications.

High-level synthesis is essentially a sequence of tasks that transforms a behavioral representation of a design into the RT level. The tasks defined by the high-level synthesis script are represented by a control data flow graph (CDFG). The CDFG illustrates data dependency between tasks and thus also identifies which tasks can be done in parallel. All tasks in high-level synthesis are broadly categorized into three classes:

1. *Allocation:* During allocation, type and quantity of resources are identified.

2. *Scheduling:* During scheduling, various operations and computations are scheduled on necessary resources. Scheduling may be a time constraint to meet the target performance, or it may be a resource constraint to meet the target area.

3. *Binding:* The binding tasks assign operations and memory accesses within each clock cycle to available hardware. It can be storage binding, such as assigning two variables to one register at different times; functional unit binding of an operation in a control step; or interconnection binding, such as assigning a multiplexer or bus for data transfer.

For high-level test synthesis, a testability analysis is performed on CDFG to identify nodes with poor controllability or poor observability or both. Based on testability analysis, various tasks are performed such as test point insertion, transformation of a register into LFSR/MISR, or conversion of a flip-flop to scan flip-flop. Some notable research in this area has been on behavioral synthesis for BIST [18], behavioral synthesis for scan [19], and design for testability synthesis at the RT level [20]. Although promising, high-level test synthesis is still in its infancy. At the present time, a robust tool for high-level test synthesis is not available in the commercial market and from the ASIC design and EDA perspective it is an open research topic.

6.6   Testing of Microprocessor Cores

While embedded microprocessors and microcontrollers are one of the key components of SoC, their testing is quite complex. This test complexity further increases in SoC due to the issues discussed in Section 6.2. In general, various design-for-test and built-in self-test schemes such as scan, partial scan, logic BIST, and scan-based BIST are used to test various logic blocks within a microprocessor/microcontroller. The main problem in these approaches is the resulting area and performance penalties. For example, full-scan implementation for an embedded microcontroller of the size about 40K gates requires an approximately 10% area overhead, and scan-based BIST requires an additional 10% to 15% area of the microcontroller. In a large number of cases, the end application of SoC is consumer electronics in which this large-area overhead is not acceptable. On the other hand, structural test methods such as scan and BIST are desirable for test reuse, portability, and even core test integration into the SoC test set.

6.6.1   Built-in Self-Test Method

This section describes a BIST-type method for the testing of microprocessors and microcontrollers. This method tests the correctness of instructions and, thus, can be considered to be a functional test method.

The general structure of a microprocessor/microcontroller is shown in Figure 6.9(a). In this figure, the instruction fetch unit obtains the opcode of the next instruction based on the address in the program counter. This opcode is decoded by the instruction decode logic, which generates function select and control signals for the execution unit. Based on these control signals, one of the logic blocks within the execution unit computes its function.

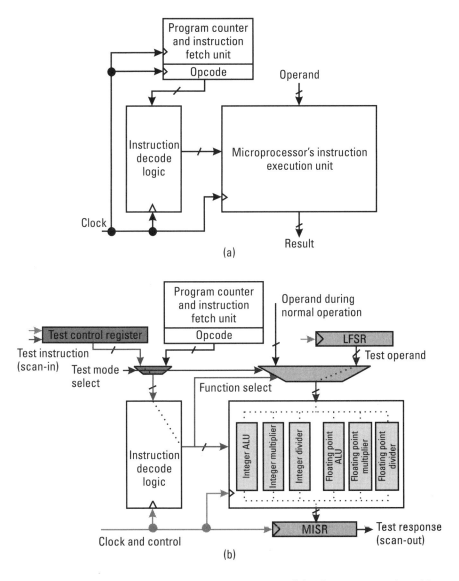

Figure 6.9 (a) General architecture of a microprocessor. (b) BIST implementation with a complex execution unit.

The operands or data for this computation are obtained from the system memory.

The structure shown in Figure 6.9(a) is modified by adding three registers: one test control register (TCR) to provide the opcode of microprocessor's instructions during the test mode, one linear feedback shift register (LFSR), and another multiple-input feedback shift register (MISR) to generate random data and to compress the test response, respectively. The data from the LFSR is used as the operand for the instruction provided by the TCR. The computed result is stored in the MISR. Note that as shown in Figure 6.9(b), the execution unit could be a simple ALU or a group of complex blocks implementing integer and floating-point arithmetic and logic operations. This modification is illustrated in Figure 6.9(b).

The sequence of testing operations in this scheme is as follows [21]:

1. Activate test mode. In this mode, the contents of the TCR are used as an instruction rather than the value from the instruction fetch unit.

2. Initialize TCR, LFSR, and MISR (either by direct control or via test control signals from the boundary-scan controller, depending on implementation).

3. Load TCR with the opcode of an instruction. Again, based on implementation, it can be either parallel load or serial load (such as scan-in).

4. Clock LFSR and MISR either for a fixed number of cycles or $2^N - 1$ cycles (full-length) for an N-bit LFSR. This step repeatedly executes the instruction in TCR with LFSR data. For example, if 1000 clocks are used, the instruction in TCR is executed 1000 times with 1000 different operands (random data provided by the LFSR).

5. Take out the content of the MISR to determine pass/fail.

6. Compare the content of the MISR with a precomputed simulation signature to determine if there is a fault. The automatic test equipment (ATE) can perform this comparison.

7. Repeat steps 2 to 6 with different instructions until all instructions are exercised.

This sequence of operations assumes that, after design completion, a simulation testbench is developed that exercises all instructions with LFSR data and MISR signatures recorded after each run. Thus, the fault-free MISR content after the each instruction run is known through simulation. Also, some instructions have loops in data flow; these instructions may require test point insertion for this method.

The above procedure determines that each instruction is executed correctly and, hence, provides functional coverage. If stuck-at fault coverage is desired, then fault simulation can be performed with various values of m (number of patterns generated by LFSR). For example, fault simulation with $m = 1000$, 10,000, or exhaustive length will provide different levels of stuck-at fault coverage. Based on this, one can select a value of m for a particular application. For fault simulation, any commercial EDA tool can be used.

The control of this scheme can be implemented in many ways as indicated in Figure 6.10. Implementation through a logic tester (ATE) is straightforward. In such a case, the tester provides the clock and test control signals and also evaluates the test response (MISR signature) to determine pass/fail.

In a case when IC has either an on-chip test controller or boundary scan, the test control signals and test response are passed through (or even controlled by) the on-chip test controller or boundary-scan TAP controller. For example, RUNBIST instruction in the boundary-scan TAP controller can be implemented to generate the test control signals; the opcode in TCR can be scanned in and the test response can be scanned out through the TAP controller. Figure 6.10 illustrates such an implementation.

The method described above is particularly useful for DSP cores, mainly because DSP cores primarily contain adder and multiplier units. This provides very little or no loop in the data flow and, thus, the execution path from the LFSR to the MISR is almost combinational during instruction.

It is also worth noting that the instruction execution through TCR permits execution of one instruction at a time, which is in a sense single-step execution, which is extremely desirable for debugging. This is further discussed in Section 6.6.3.

6.6.2 Example: Testability Features of ARM Processor Core

Sections 6.2 and 6.3 contain all the essential elements for the testing of digital logic cores. Although the elements discussed in Sections 6.2 and 6.3 are under discussion by the P1500 group, they have not yet been approved as a standard, so some of these concepts as well as some other approaches have been used by the core providers. In this section, we provide an example of a popular ARM processor to illustrate the mix test approach in use by the core provider.

ARM Limited has a number of processor cores, descriptions of which are available on ARM's web site (see list in Chapter 1). From an architectural point of view, these processor cores can be characterized as follows:

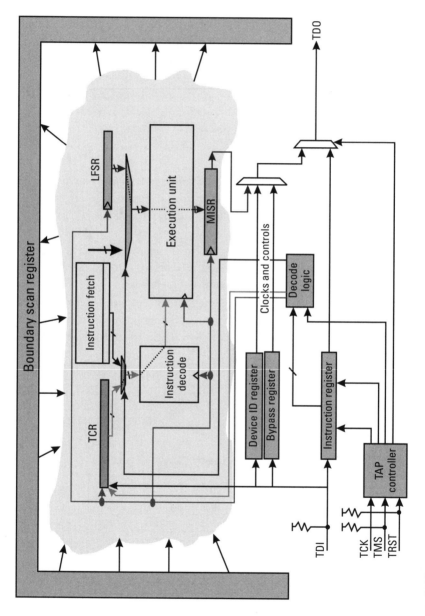

Figure 6.10 BIST implementation through TAP controller.

- *ARM7TDMI:* 32-bit RISC CPU, thumb instruction set extension, embedded ICE debugging, and a DSP-enhanced multiplier;
- *ARM740T, ARM710T, and ARM720T:* Cached variant based on ARM7TDMI core;
- *ARM9TDMI:* 5-stage pipeline and Harvard buses, more than twice the performance of ARM7TDMI;
- *ARM940T and ARM910T:* Cached variants based on ARM9TDMI core;
- *StrongARM:* High-performance cached processor.

ARM Limited provides a C language model of the core to protect their IP during RTL development of SoC. ARM also has tools to wrap the C model for a variety of Verilog/VHDL simulators. Once front-end design is complete, an ARM hard macrocell is incorporated into the netlist of the SoC during layout. The GDSII of the ARM hard macro is available for a number of process technologies and fabs who have partnership relationships with ARM Limited.

ARM also provides three test approaches, which the SoC integrator needs to combine with the test methods of other blocks.

1. *Parallel test:* This is extracted from the functional simulation. The functional simulation uses a combination of code and testbenches like any other design. In general, it can be considered to be the testbench for instruction verification.

2. *Serial test:* This utilizes a boundary scan at the I/Os of the ARM core, which contains 105 elements on the scan chain. For this, the parallel vectors are converted into serial format.

3. *Test interface controller (TIC):* This is a bus-based approach that uses parallel vectors converted into TIC format. This approach requires that the peripheral cores have a wrapper through which inputs can be applied and output can be observed. With this, each peripheral becomes testable using bus accesses, which can be generated by any element in the system such as TIC. The concept is illustrated in Figure 6.11 [22].

The TIC-based testing is done using 32-bit vectors (the width of the bus). For an ARM core, four vectors are required for each processor clock. Thus, a simple state machine is used to control the application of vectors.

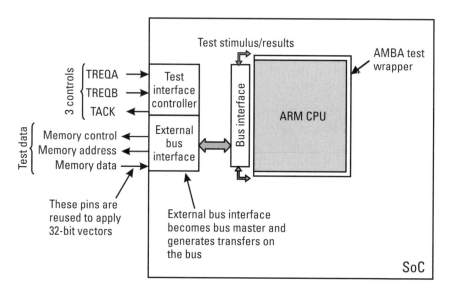

Figure 6.11 Illustration of ARM-based SoC testing using a test interface controller. (From [22], © IEEE 1999. Reproduced with permission.)

Also, a processor test wrapper is used that contains a register to store control signals and provide muxing at outputs for observation.

The TIC file format is very similar to the C language in its support of variables, loops, and function calls. From the TIC file a test intermediate file (TIF) is created in ASCII format to help users understand the simulation environment. The test vectors are actual simulation vectors and a subset is generally used during the production test.

6.6.3   Debug Support for Microprocessor Cores

In a processor, many things can happen that never show up at the I/Os and thus debugging is often a major challenge. To overcome this problem, breakpoints in hardware and software are asserted. These breakpoints consist of address compare functions on the instruction and data addresses, and data compare functions. The basic objective is to observe all or any desired internal states at the breakpoint. The maximum breakpoint assertion is equivalent to single-step execution in which one instruction at a time is executed and internal states are observed.

ARM uses the popular concept of an in-circuit emulator (ICE) for debug support. The embedded ICE module is simply the user-accessible portion of the debugger module. ARM provides two options (called option D and option I for data and instruction, respectively), typically, both options are designed together into a custom part. Because embedded ICE uses considerably more logic than the debugger itself, it is generally removed at the production stage without disturbing the processor and debugger.

The debug module of the ARM processor provides breakpoint (instruction break) and watchpoint (data break) registers as well as a serial data interface. The processor enters into the debug state as a result of a breakpoint assertion or watchpoint hit. In the debug state, the instructions and control information are loaded directly into the processor, and status information is read through the serial interface. This debug state can also be forced onto the processor by an input signal (this behavior is equivalent to an interrupt). The debug module also adds two scan chains at the ARM processor I/Os. These scan chains operate like a boundary scan under a protocol similar to IEEE 1149.1. The larger scan chain is generally used to monitor the status of all external signals (from the ARM processor's view), while the smaller chain is generally used to sample the data bus.

Many of the debugging issues can be simplified by providing simple provisions in the design. For example, in principle, by implementing only one extra PIO register in the microprocessor core such that it can be scanned in and scanned out through the boundary scan, the contents of any internal register can be observed, and any test data/address can be loaded onto any of the internal registers. In other words, one instruction can be executed followed by data transfer between the desired register and the scannable register, and scanning out data from the scannable register to observe the content of the desired register.

Based on this concept, MIPS Technology together with Philips, Toshiba, LSI Logic, IDT, Sony, NEC, NKK, and QED have developed the Enhanced JTAG (EJTAG) method for internal use to facilitate debugging. The complete specifications are given in [23], which also includes a detailed description of the EJTAG interface with an MIPS CPU core.

In SoC, further complication in debugging occurs when the design contains both a microprocessor core and a DSP core. Because of the different architectures, good observability of the interface is required. In such designs, both cores generally access external memory through a common address and data bus. Thus, during the debugging of one core, other cores stall, hiding any adverse effect on them. Because of these problems, in dual (or multiple) core designs, boundary-scan access with TAP is preferred for all cores. Ideally, the user should be able to work with both debuggers as if they were talking to discrete devices, uncovering the hidden interface between the cores.

Examples of such situations are Motorola's Redcap product, which combines an M-core RISC with a DSP56600 core; and GEC's GEM301 global system for mobile communication, which contains an ARM7TDMI processor with DSP Group's Oak DSP core.

References

[1] D&T Roundtable, "Testing embedded cores," *IEEE Design and Test of Computers,* April–June 1997, pp. 81–89.

[2] Rajsuman, R., "Challenge of the 90's: Testing CoreWare based ASICs," Panel on DFT for Embedded Cores, *Proc. IEEE Int. Test Conf.,* 1996, p. 940.

[3] Hutcheson, J., "Executive advisory: the market for systems-on-a-chip," June 15, 1998, and "The market for systems-on-a-chip testing," July 27, 1998, VLSI Research, Inc.

[4] IEEE P1500 web page, http://www.grouper.ieee/group/1500, and "Preliminary outline of the IEEE P1500 scalable architecture for testing embedded cores," IEEE VLSI Test Symposium, 1999, pp. 483-488.

[5] Zorian, Y., "Test requirements for embedded core based systems and IEEE P1500," *Proc. IEEE Int. Test Conf.,* 1997, pp. 191–199.

[6] Zorian, Y., E. J. Marinissen, and S. Dey, "Testing embedded core based system chips," *Proc. IEEE Int. Test Conf.,* 1998, pp. 130–143.

[7] Marinissen, E. J., et al., "Towards a standard for embedded core test: An example," *Proc. IEEE Int. Test Conf.,* 1999, pp. 616–627.

[8] IEEE Standard 1450, *STIL,* 1999.

[9] IEEE Standard 1149.1, *IEEE Standard Test Access Port and Boundary Scan Architecture,* IEEE Press, 1990.

[10] Whetsel, L., "An IEEE 1149.1 based test access architecture for ICs with embedded cores," *Proc. IEEE Int. Test Conf.,* 1997, pp. 69–78.

[11] Bhattacharya, D., "Hierarchical test access architecture for embedded cores in an integrated circuit," *Proc. IEEE VLSI Test Symp.,* 1998, pp. 8–14.

[12] Hodor, K., "Protecting your intellectual property from pirates," *DesignCon 98.*

[13] Lach, J., W. H. M. Smith, and M. Potkonjak, "Signature hiding techniques for FPGA intellectual property protection," *Proc. IEEE Int. Conf. on CAD,* 1998, pp. 186–188.

[14] Kirovski, D., et al., "Intellectual property protection by watermarking combinational logic synthesis solutions," *Proc. IEEE Int. Conf on CAD,* 1998, pp. 194–198.

[15] De, K., "Test methodology for embedded cores which protects intellectual property," *Proc. IEEE VLSI Test Symp.,* 1997, pp. 2–9.

[16] Touba, N. A., and B. Pouya, "Using partial isolation rings to test core based designs," *IEEE Design and Test of Computers,* Oct.–Dec. 1997, pp. 52–59.

[17] Pouya, B., and N. A. Touba, "Modifying user-defined logic for test access to embedded cores," *Proc. IEEE Int. Test Conf.,* 1997, pp. 60–68.

[18] Avra, L. J., and E. J. McCluskey, "High level synthesis of testable designs," Test Synthesis Seminar, *Proc. IEEE Int. Test Conf.,* 1994.

[19] Potkonjak, M., S. Dey, and R. K. Roy, "Behavioral synthesis of area efficient testable design using interconnection between hardware sharing and partial scan," *IEEE Trans. CAD,* Sep. 1995, pp. 1141–1154.

[20] Bhattacharya, S., F. Bgrlez, and S. Dey, "Transformation and resynthesis for testability of RT-Level control data path specification," *IEEE Trans. VLSI,* Sep. 1993, pp. 304–318.

[21] Rajsuman, R., "Testing a system-on-a-chip with embedded microprocessor," *Proc. Int. Test Conf.,* 1999, pp. 499–508.

[22] Harrod, P., "Testing reusable IP—A case study," *Proc. IEEE Int. Test Conf.,* 1999, pp. 493–498.

[23] MIPS EJTAG Debug Solution, 2.0, MIPS Technology Inc., 1998.

7

Testing of Embedded Memories

Embedded memories are essential components in SoC designs. Embedded memories have increased in size dramatically during the past few years, as observed by various industry observers as well as by the ITRS road map [1]. As shown in Figure 1.5, SoCs may contain multiple memories, some of which may be multimegabits in size. A number of companies provide SRAM, DRAM, ROM, EEPROM, and flash memory compilers and modules that are optimized to a specific process technology. For example, Artisan's SRAM and DRAM blocks and Virage Logic's EEPROM and flash compilers are quite popular. Many ASIC companies have also developed in-house memory compilers as well as partnerships with the memory vendors to develop multimegabit DRAM blocks for embedded use. These embedded memories implement register files, FIFOs, data cache, instruction cache, transmit/receive buffers, storage for audio/video, graphics texture processing, and so on.

Testing of embedded memories is one of the major problems in today's embedded core-based SoCs as well as in complex microprocessors. In a significant number of cases, these memories are tested using built-in self-test methods. However, a number of methods are used in SoC designs to test these memories. The testing of embedded memories is not a new topic. Because memories are used as a test vehicle for technology development, memory testing is a fairly well-studied topic. In addition to testing, memory repair in SoC is also a key issue for large embedded memories. In this chapter, we discuss both testing and repair of embedded memories.

155

7.1   Memory Fault Models and Test Algorithms

Memory fault models are significantly different than the fault models used for digital logic. Line stuck-at faults, bridging, opens, and transistor stuck-on/-off fault models work fairly well for digital logic [2], however, these fault models are insufficient for determining the functional correctness of memories. In addition to line stuck-at, bridging, and open faults, memory faults also include bit pattern, transition, and cell coupling faults.

7.1.1   Fault Models

The commonly used faults in memories are as follows:

- *Line stuck-at faults:* Includes single and multiple lines (input, output, address, or bit line) stuck at logic "0" or logic "1."

- *Cell stuck-at faults:* Refers to a memory cell stuck at either "0" or at "1."

- *Bridging faults:* Includes single and multiple bridging faults. For simplicity, low-resistance bridging is considered in the majority of cases; however, actual defects also result in high-resistance bridging. Bridging may occur among input lines, output lines, address lines, bit lines, or among a combination of input, output, address, and bit lines. Bridging among inputs and outputs is rarely considered, because their probability of occurrence is low.

- *Open faults:* Includes single and multiple opens at input, address, bit, or output lines.

- *Addressing faults:* A row or column decoder may access the addressed cell (d^- fault) or a nonaddressed cell (d^+ fault), or it may access multiple cells. It is also possible that the decoder will not access the specified cell but some other cell ($d^- + d^+$ fault).

- *Missing or extra bits:* An extra bit location may exist or a bit location may be missing from an intended location. These faults are particularly important for ROMs and are sometimes known as growth or shrinkage faults.

- *Bit-pattern faults:* In programmable ROMs such as mask programmable ROMs, EEPROMs, and flash memories, programming faults may cause an error. For example, a fusible-link-based ROM may have an unblown (or partially blown) fuse at an address location as well as a blown fuse at an unintended location. This results in a "0" instead of a "1" and vice versa. In EEPROM and flash, such bit-pattern faults occur mainly due to programming error.

- *State transition faults:* In random access memories (RAMs), state transition faults refer to transitions in the cell's data from 1 to 0 and 0 to 1.

- *Cell coupling faults:* These faults are generally considered for RAMs. Cell coupling faults mean that a specified memory location is affected (either data or transitions at a location) because of other locations (by data at other locations or transitions at other locations). Cell coupling may be the inversion type (content of cell inverts), idempotent type (content of cell changes only if cell has a specific data), or simple state coupling (content of cell changes only by a specific data at other locations).

- *Data retention:* This refers to a fault that occurs because the memory cell could not hold data for a specified period of time. Data retention faults are very important for RAMs and sometimes for programmable ROMs and flash memories.

- *Data endurance:* In EEPROMs and flash memories, the amount of charge is reduced after multiple read/write operations. Subsequently, an error may occur in reading the right data.

- *Pattern-sensitive faults (PSF):* This is a special case of state coupling faults. It means that in the presence of some specific data in one part of memory, data in some other part of memory may be affected. Pattern-sensitive faults may be dynamic (due to change in data) or static (due to fixed data) as well as global or only limited to neighborhood cells.

In addition to these faults, parametric and timing faults are also considered. Timing faults include data access time, which is a very important parameter for memories. Furthermore, transistor stuck-on/-off faults are sometimes also considered.

7.1.2 Test Algorithms

A large number of test algorithms have been reported in the literature [3–5]. A few commonly used algorithms are included in this section.

- *Memory scan or MSCAN:* This is the simplest test. First a "0" is written to each cell, the value is verified, and a "1" is written and verified. This procedure is very fast, but its fault coverage is limited. For *n*-bit memory, the algorithm can be given as follows:

```
For I = 1 to n; Do
  Write 0 to cell I
  Read cell I for 0
  Write 1 to cell I
  Read cell I for 1
Continue
```

- *Checker pattern:* In this test, a pattern of alternate 0s and 1s is written to memory array. After a pause, the whole array is read. In a modified version, additional read/write operations are done using a complementary pattern. This is also a quick test that provides cell stuck-at faults, data retention, and 50% state transition faults.

- *Galloping pattern (GALPAT):* In this test, the entire memory array is ‹ first initialized to 0. Then 1 is written to a cell in ascending order of addresses and the whole array is read back. The algorithm is shown in Figure 7.1(a). The complexity of this algorithm is $O(n^2)$; however, it also provides extensive fault coverage including cell stuck-at, state transition, and a large number of coupling and pattern-sensitive faults.

- *Galloping diagonal/column (GALDIA/GALCOL):* These are modified versions of GALPAT such that instead of writing a 1 in a cell, 1 is written in the whole diagonal/column. This speeds up the testing because the number of read/write operations is reduced to $O(n^{3/2})$. But at the same time, coverage for PSF and coupling faults reduces.

- *March algorithm:* This is one of the most popular memory test algorithms. In this test, a sequence of operations is performed on one cell before proceeding to the next cell. This sequence of operations is called the "March element." A March element may constitute a simple sequence of MSCAN or it may constitute a complex sequence with many writes and reads. The basic March algorithm (also known as MATS) is illustrated in Figure 7.1(b). Because the read/write operations are done on an individual cell, the complexity of read/write operations in this algorithm is $O(mn)$, where m is the number of read/write operations in the March element. The fault

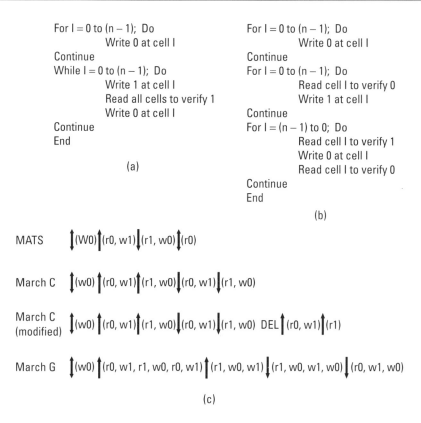

For I = 0 to (n − 1); Do
    Write 0 at cell I
Continue
While I = 0 to (n − 1); Do
    Write 1 at cell I
    Read all cells to verify 1
    Write 0 at cell I
Continue
End

(a)

For I = 0 to (n − 1); Do
    Write 0 at cell I
Continue
For I = 0 to (n − 1); Do
    Read cell I to verify 0
    Write 1 at cell I
Continue
For I = (n − 1) to 0; Do
    Read cell I to verify 1
    Write 0 at cell I
    Read cell I to verify 0
Continue
End

(b)

MATS $\updownarrow$(W0)$\uparrow$(r0, w1)$\downarrow$(r1, w0)$\updownarrow$(r0)

March C $\updownarrow$(w0)$\uparrow$(r0, w1)$\uparrow$(r1, w0)$\downarrow$(r0, w1)$\downarrow$(r1, w0)

March C (modified) $\updownarrow$(w0)$\uparrow$(r0, w1)$\uparrow$(r1, w0)$\downarrow$(r0, w1)$\downarrow$(r1, w0) DEL$\uparrow$(r0, w1)$\uparrow$(r1)

March G $\updownarrow$(w0)$\uparrow$(r0, w1, r1, w0, r0, w1)$\uparrow$(r1, w0, w1)$\downarrow$(r1, w0, w1, w0)$\downarrow$(r0, w1, w0)

(c)

Figure 7.1 (a) GALPAT test algorithm. (b) March test algorithm. (c) A few examples of various March tests. Order of addressing: Up arrows represent ascending order, down arrows represent descending order, and double-headed arrows represent any order. DEL represents a delay or pause.

coverage of this test is quite good, which includes cell stuck-at, state transition, and a large number of PSF and coupling faults.

Many variations on the March algorithm have been reported in the literature [3–8] that have basically been developed using different March elements. One simple modification involves the memory being disabled for some time using a delay element upon completion of the test, and a few additional read/write operations being performed. The purpose of this delay element is to detect data retention faults. Some selected variations are illustrated in Figure 7.1(c). The most popular March algorithms are MATS, which has complexity $O(6n)$, and March C, which has complexity $O(9n)$.

7.1.3   Effectiveness of Test Algorithms

A number of studies have been conducted to determine which test algorithm is best at detecting failures in actual memory ICs [7, 8]. The summary of results is illustrated in Figure 7.2. It is interesting to note that some of the time-consuming test algorithms such as shifted diagonal and galloping column provide poor fault coverage and hence do not justify their lengthy test times. On the other hand, the March algorithm provides excellent fault coverage. Note, however, that some of its lengthy and complex variations do not provide significant improvement in fault coverage.

In general, none of the algorithms are restricted to a particular architecture; however, different algorithms are limited to different kinds of faults. A very interesting observation can be made with respect to surround disturb and API tests that target pattern-sensitive faults. These algorithms provide poor coverage of actual failures; hence, it can be concluded that PSF is not a good fault model. On the other hand, algorithms that target coupling faults provide excellent coverage of physical defects and actual failures and, hence, the coupling fault can be considered to be a good fault model.

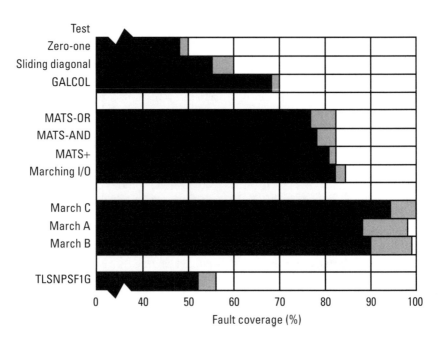

Figure 7.2   Comparison of various memory test algorithms. (From [6], © IEEE 1993. Reproduced with permission.)

7.1.4 Modification With Multiple Data Background

All the algorithms discussed in Section 7.1.2 are based on bit-oriented read/write operations. Almost all embedded memories in SoC are word oriented. Also, to reduce the test time, large DRAMs and flash memories are designed such that they appear to be byte oriented during the test mode. In general, if a bit-oriented memory is tested as an M-bit-wide word-oriented memory, $(K+1)$ different data backgrounds should be used during testing, where $K = \log_2 M$. For example, for byte-oriented memory (word size being 8 bits), there are eight possible data backgrounds (01010101; 10101010; 11001100; 00110011; 11110000; 00001111; 11111111; 00000000), however, only four backgrounds are necessary (10101010; 11001100; 11110000; 11111111). Thus, the single-bit March algorithm should be replicated four times with the following replacements:

1. In the first replica, w0 and w1 by w01010101 and w10101010;
2. In the second replica, w0 and w1 by w00110011 and w11001100;
3. In the third replica, w0 and w1 by w00001111 and w11110000;
4. In the fourth replica, w0 and w1 by w00000000 and w11111111.

The r0 and r1 operations are also replaced by the same rule. Thus, by using an M-bit-wide word for read/write operations and a different data background, the test time changes by a factor of $(M+1)/\log_2 M$.

7.1.5 Modification for Multiport Memories

Single-port memories have limited bandwidth for data access. In a large number of SoCs designed for audio/video and graphics applications, simultaneous read and write from different ports is required. For multiport memory, each read/write port should have sequences $r(x)w(\overline{x})$ and $r(\overline{x})w(x)$ in both increasing and decreasing addresses in order to cover address decoder faults [9]. Also, the read operation is done on all read ports, whereas the write operation is done on one write port at a time. These ports are called *active ports*. The inactive ports are addressed such that they do not influence the active read and write ports. The inactive ports address a fixed address, such as address 0. An example for 3-write and 6-read port memories is shown in Table 7.1.

7.1.6 Algorithm for Double-Buffered Memories

Double-buffered memories (DBMs) using master and slave parts for data buffering are also commonly used in SoC designs, particularly for implementing

Table 7.1

Extension of March Algorithm for 3W/6R Memory

Addressing	Pass 1	Pass 2	Pass 3	Pass 4	Pass 5	Pass 6	Pass 7
	W(0)	R(0)W(1)	R(1)W(0)	R(0)W(1)	...	...	R(1)W(0)
	...	...	...	...	...	...	...
	W(0)	R(0)W(1)	R(1)W(0)	R(0)W(1)	...	...	R(1)W(0)
Active read ports	All	All	All	All	All	All	All
Active write ports	Any	A	A	B	B	C	C

Note: Different write ports are represented by A, B, and C.

first-in/first-out (FIFO) stacks. This conditional buffering is controlled by a global transfer signal, which transfers data from the master to the slave. Thus, each DBM cell contains two locations that behave in a master/slave con- figuration. Also, a shift register is often placed on top of the memory array to facilitate pointer addressing. From a testing point of view, the buffering mechanism and pointer addressing in DBMs limit a cell's observability and controllability.

To simplify DBM testing, the DBM is divided into three substructures: the memory array, the read/write logic, and the address generation and decoder logic. The faults are considered in both master and slave locations.

The test algorithm for DBM with NW words of NB bits is shown in Figure. 7.3 [10]. In Figure 7.3, 0r and 1r represent all-0 and all-1 words. A write operation is defined as $[Wr(\infty)]_i$; a read operation is defined as $[Rd(\infty)]_i$ for a word with address i, where ∞ means all-0 or all-1 word. The master and slave are denoted by $[M]_i$ and $[S]_i$, respectively, and the value contained by the master and slave are denoted by $V([M]_i)$ and $V([S]_i)$ respectively. The test length of the algorithm shown in Figure 7.3, is 53N for bit-oriented memory. For word-oriented memory, the algorithm can be modified as explained in Section 7.1.4.

7.2   Test Methods for Embedded Memories

Many methods for test vector application and response evaluation are available for embedded memories. However, each method has its own advantages and disadvantages. In this section, we briefly describe these methods.

1. $[Wr(0r)_i$, for $i = 1, \ldots NW$
2. T
3. $[Rd(0r)]_i$ & $[Wr(1r)_i$, for $i = 1, \ldots NW$
4. T
5. $[Rd(1r)]_i$ & $[Wr(1r)_i$, for $i = 1, \ldots NW$
6. T
7. $[Rd(0r)]_i$ & $[Wr(0r)_i$, for $i = 1, \ldots NW$
8. T
9. $[Rd(0r)]_i$ & $[Wr(0r)_i$, for $i = 1, \ldots NW$
10. T
11. $[Rd(0r)]_i$ & $[Wr(1r)_i$; $[Rd(0r)]_i$; T, for $i = 1, \ldots NW$
12. $[Rd(1r)]_i$ & $[Wr(0r)_i$; $[Wr(1r)]_i$; T, for $i = 1, \ldots NW$
13. $[Rd(1r)]_i$ & $[Wr(0r)_i$; $[Wr(1r)]_i$; $[Rd(1r)]_i$; T, for $i = 1, \ldots NW$
14. $[Rd(1r)]_i$ & $[Wr(0r)_i$; $[Rd(1r)]_i$; T, for $i = 1, \ldots NW$
15. $[Rd(0r)]_i$ & $[Wr(1r)_i$; $[Wr(0r)]_i$; T, for $i = 1, \ldots NW$
16. $[Rd(0r)]_i$ & $[Wr(1r)_i$; $[Wr(0r)]_i$; $[Rd(0r)]_i$; T, for $i = 1, \ldots NW$
17. $[Rd(0r)]_i$ & $[Wr(1r)_i$, for $i = NW, \ldots 1$
18. T
19. $[Rd(1r)]_i$ & $[Wr(1r)_i$, for $i = NW, \ldots 1$
20. T
21. $[Rd(1r)]_i$ & $[Wr(0r)_i$, for $i = NW, \ldots 1$
22. T
23. $[Rd(0r)]_i$ & $[Wr(0r)_i$, for $i = NW, \ldots 1$
24. T
25. $[Rd(0r)]_i$ & $[Wr(1r)_i$; $[Rd(0r)]_i$; T, for $i = NW, \ldots 1$
26. $[Rd(0r)]_i$ & $[Wr(1r)_i$; $[Wr(1r)]_i$; T, for $i = NW, \ldots 1$
27. $[Rd(1r)]_i$ & $[Wr(0r)_i$; $[Wr(1r)]_i$; $[Rd(1r)]_i$; T, for $i = NW, \ldots 1$
28. $[Rd(1r)]_i$ & $[Wr(0r)_i$; $[Rd(1r)]_i$; T, for $i = NW, \ldots 1$
29. $[Rd(0r)]_i$ & $[Wr(1r)_i$; $[Wr(0r)]_i$; T, for $i = NW, \ldots 1$
30. $[Rd(0r)]_i$ & $[Wr(1r)_i$; $[Wr(0r)]_i$; $[Rd(0r)]_i$; T, for $i = NW, \ldots 1$

Figure 7.3 Test algorithm for double-buffered memory. (From [10], © IEEE 1993. Reproduced with permission.)

7.2.1 Testing Through ASIC Functional Test

For small memories, ASIC vendors include simple read/write operations in the ASIC functional test. Most of the time the 1010…10 pattern and its reverse is written and then read. Generally, this method is applicable only to small memories and extensive testing is not done by this method.

7.2.2 Test Application by Direct Access

Direct application of test while accessing memory through I/O muxing has been widely used for embedded memories. This concept is illustrated in Figure 7.4. This method requires modification in the I/Os by adding a multiplexer. Due to this extra mux, there is a permanent penalty in the chip's performance. In this method, the tester's ALPG unit generates the test patterns. However, due to the mux at the I/Os, actual patterns require serialization of ALPG patterns, which increases test complexity, test time, and often loss of at-speed testing.

7.2.3 Test Application by Scan or Collar Register

Test application to memory through local boundary scan or a collar register is generally used for small embedded memories. This concept is illustrated in Figure 7.5. This method adds a wrapper (boundary-scan or shift-register-type wrapper) to embedded memory. Thus, the data transfer rate to and from memory slows down by the time equal to the delay of the wrapper. During testing, the test patterns are serially shifted in and response is serially shifted out. Thus, the test time increases significantly and at-speed testing is not possible.

7.2.4 Memory Built-in Self-Test

In the past few years, built-in self-test has been recognized as an extremely important methodology for embedded memories. While RAMBIST

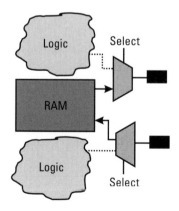

Figure 7.4 Illustration of direct access to embedded memory through multiplexing.

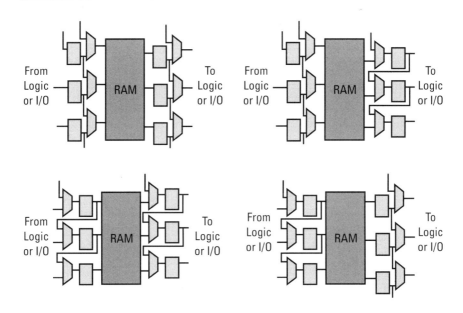

Figure 7.5 Testing of embedded memories using a wrapper or scan chain. Many other configurations are possible besides the four configurations shown here.

methods have drawn much attention, BIST methods for ROM, EPROM, and flash are also on the rise.

BIST methods require an additional circuit for on-chip test generation and response evaluation. A general RAMBIST architecture is shown in Figure 7.6 [9]. In general, an EDA tool creates file(s) that contain all the additional circuitry. In this architecture, the address generator is generally a counter, the self-test controller is a finite-state machine, and the data receiver is an XOR tree. As shown in Figure 7.6, the test circuitry itself can be tested through scan methodology. The area overhead of self-test controller, address/data generator is about 500 to 1000 gates. However, due to XOR tree, the area overhead of the data receiver can be very large particularly if the memory word size is large. Also, when word size is large, data-in/data-out require a large number of wires to be routed. An example of a SoC using this method while the RAMBIST controller is shared by two large embedded RAMs is illustrated in Figure 7.7 [9].

To simplify the routing of the method illustrated in Figure 7.6, a few EDA vendors have modified this architecture by serializing data-in/data-out. One such modification was developed at Northern Telecom for synchronous RAMs and licensed by Logic Vision [11]. This structure is shown in

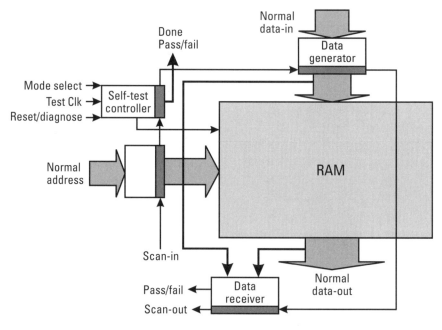

Figure 7.6 General architecture of RAMBIST. (From [9], © IEEE 1996. Reproduced with permission.)

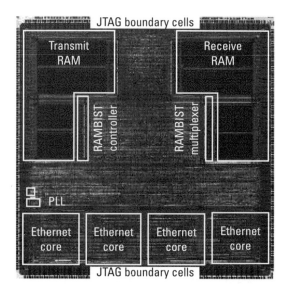

Figure 7.7 Example of an Ethernet controller SoC with two large embedded RAMs tested using a shared RAMBIST controller. (From [9], © IEEE 1996. Reproduced with permission.)

Figure 7.8(a). To serialize the data, muxes are inserted between RAM latches. With these muxes, the synchronous latches form a scan-chain-type structure, in which 1 bit of test data is fed. As 1 bit is fed every cycle and bits from previous cycles are latched into memory, read/write operations similar to the bit-oriented March algorithm take place. The nth bit data-out is returned back to the controller, where 1-bit XOR operation determines pass/fail. Note that because of the 1-bit pass/fail operation, the comparator status signal requires continuous monitoring to identify a failed word (row). The specific failed bit location cannot be identified.

To know the result of a cycle-by-cycle comparison, the compare status signal requires continuous monitoring and thus needs to be the primary output. If the user wishes to enhance the failure resolution, multiple bits (such as $n/4$, $n/2$, $3n/4$, and nth) can be tapped and sent to the controller and compared to identify the failure. However, this requires multiple compare status signals. In such a case, the failed segment of the word is identified. If all bits are compared and the compare status becomes n bits wide, then a failed bit can be identified. In that case, the comparator becomes XOR-tree, similar to Figure 7.6, and memory data outputs become primary outputs, similar to direct access testing.

To avoid n-bit XOR tree and n wires for the comparator, a different enhancement was developed by Heuristic Physics Lab (HPL) (that segment of HPL has now been acquired by Credence Systems). The scheme is somewhat similar to that shown in Figure 7.8(a). However, instead of serializing test data through 1 wire, m wires were used ($m \ll$ word size). These m wires are decoded at the boundary of memory to create n-bit test data, which is launched to the memory. Similarly, data-out from memory is encoded into m bits at the boundary of memory and this encoded data is sent to the comparator. This scheme is illustrated in Figure 7.8(b). It uses a little bit more area and routing than the LogicVision/Northern Telecom method; however, it identifies the failed bit location. It does not require any compare status signal(s), and it does not require continuous monitoring of any compare signals.

Compared to other test methods, BIST is the costliest in terms of hardware overhead. The commercially available memory built-in self-test methods require a memory BIST controller and some additional circuitry at the periphery of memory. The total area overhead in RAMBIST is about 3% to 4% for 16-Kbit memory. Also, due to additional circuit parasitics about a 1% to 2% performance penalty occurs in memory read/write operation.

Because memory BIST provides deterministic testing in acceptable test times, little performance penalty at the chip's I/Os, and only about a 1% to

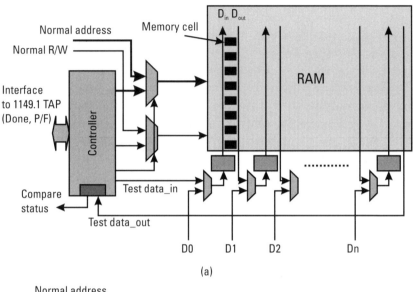

(a)

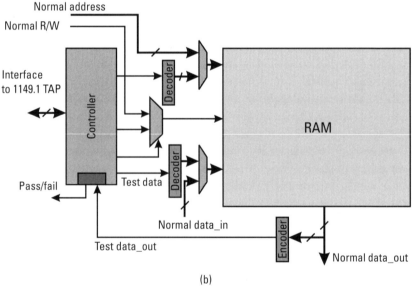

(b)

Figure 7.8 Memory BIST (a) with serial interface to reduce routing and XOR-tree comparator and (b) with data decoding/encoding at memory boundary to simplify routing.

2% penalty in memory read/write operations, it is increasingly used for embedded memories in SoC.

A number of memory BIST methods are available commercially from vendors like Mentor Graphics and LogicVision. A number of semiconductor companies have also developed in-house CAD tools for memory BIST. References [9, 11–15] describe some of the memory BIST methods. References [9] and [16] also describe example CAD tools for implementing memory BIST.

7.2.5 Testing by On-Chip Microprocessor

In this method, the on-chip microprocessor acts as a tester. An assembly language program in the microprocessor mnemonics is used to implement the desired memory test algorithm. The executable is obtained through the microprocessor's assembler. This executable allows the microprocessor to behave as a test generator and as an evaluator during testing [17].

In this method, the computational power of this microprocessor core is utilized to generate the memory test pattern, apply the test patterns, and evaluate the test response. For this purpose, an assembly language program is executed at the microprocessor core to generate memory test patterns. An example of the March algorithm is shown in Figure 7.9; however, the methodology is not limited to it and any algorithm can be used. The test response can be evaluated by the microprocessor or by a simple comparator circuit. The procedure shown in Figure 7.9 is based on a March pattern in word mode with 0101…01 data (5555H) in increasing order and 1010…10 data (AAAAH) in decreasing order for 16Kx16 RAM.

Simplified forms of such procedures have been used by a number of companies [18–20]. At the present time, the BIST or direct access method is used to test one of the on-chip memories—for example, the instruction cache in a microprocessor. This known good memory is then used to store the executable test program (similar to the program given in Figure 7.9). The execution unit of the microprocessor takes this executable test program from the instruction cache and performs testing for other on-chip memories such as data cache, z-buffer, and register files.

Another possibility is to generate executable binary from the microprocessor's assembler residing on a host computer. This executable binary can then be fed to the microprocessor from the tester through an interface API, thus eliminating the need to have one known good memory on chip.

The major advantage of this method is that no additional test circuitry is required. It does not require any design modification in the chip design, there is no performance penalty, and memory is tested at speed. The user can use any memory test algorithm to test embedded SRAM, DRAM, or any

```
/* Procedure to test embedded RAM */
/* A0 is address counter, D0 contains test data (write), D1 is used for read data (response) */
/* Register Initialization */
                    MOVE 0000H, A0          ;Initializing address counter
                    MOVE 0000H, D0          ;This is data for memory initialization
                    MOVE 0000H, D2          ;This is used to clear memory word

/* Test Procedure Initialization */
Initial             MOVE D0, [A0]
                    COMP A0, FFFFH          ;FFFFH is the last address
                    BEQ Test_Incr
                    BRA Initial             ;Memory is initialized
                    INCR A0

/* Write/Read in increasing order */
Test_Incr           MOVE 0000H, A0
                    MOVE 5555H, D0
Cont_Incr           MOVE D0, [A0]           ; This is write operation
                    MOVE [A0], D1           ;This is read operation
                    MOVE D2, [A0]           ; Clears memory word
                    COMP D0, D1
                    BEQ Next_Incr
                    BRA Fail                ; Read data is not 5555H
Next_Incr           COMP A0, FFFFH
                    BEQ Test_Decr
                    BRA Cont_Incr
                    INCR A0

/* Write/Read in decreasing order */
Test_Decr           MOVE AAAAH, D0
Cont_Decr           MOVE D0, [A0]           ;This is write operation
                    MOVE [A0], D1           ;This is read operation
                    MOVE D2, [A0]           ;Clears memory word
                    C OMP D0, D1
                    BEQ Next_Decr
                    BRA Fail                ;Read data is not AAAAH
Next_Decr           COMP A0, 0000H          ;0000H is the last address
                    BEQ Done
                    BRA Cont_Decr
                    DECR A0

Done                Write Test_Passed
Fail                Write Test_Failed
```

Figure 7.9 A word-wide March pattern generated by a microprocessor core to test 16Kx16 embedded memory with 5555H for increasing order and AAAAH for decreasing order.

other kind of memory. The method also provides full fault diagnosis (failed bit location) without any extra hardware. Due to zero hardware overhead, zero performance penalty, no modification in the design, flexibility in the test algorithm, and at-speed testing, this method is very well suited for use in SoC.

7.2.6 Summary of Test Methods for Embedded Memories

In principle, any test method can be used for any embedded memory. However, significant pros and cons are associated with each method, as summarized in Table 7.2 [17].

A possible guideline for selecting a test method based on area overhead and test generation effort is given in Table 7.3. However, it is strongly advised that every SoC design and every individual memory be analyzed carefully before selecting a particular test methodology for it.

7.3 Memory Redundancy and Repair

The primary reason for built-in self-repair (BISR) is the enhancement of memory yield and subsequently overall SoC yield. Various studies have shown that embedded memory yield can be enhanced in the range of 5% to 20% depending on the redundancy and BISR method. Because of this increased embedded memory yield, the net increase in SoC yield can be 1% or 2% to close to 10%. As an example, the yield impact on UltraSparc is shown in Figure 7.10 [21].

Generally, redundancy analysis and repair procedures are interlinked. During testing, when failures are identified and located, a redundancy analysis procedure is executed to determine which failures can be repaired using the given redundant rows and columns. Based on this redundancy analysis, a repair procedure is executed that can be either a hard repair or soft repair. An example flowchart for redundancy analysis and repair is given in Figure 7.11 [22].

7.3.1 Hard Repair

In general, hard repair uses fuses/antifuses/laser programming to disconnect rows/columns with faulty bits and replaces them with redundant rows/columns [23, 24]. This method has been used by memory vendors for some time, and that knowledge has been adopted by ASIC and SoC vendors. This

Table 7.2
Comparison of Various Test Methods for Embedded Memories

Test Method	Pros	Cons
Direct access	1. Detailed testing is possible 2. Well-established memory ALPG can be used 3. ATE fault diagnosis tools can be used	1. It requires pin multiplexing 2. It requires serialization of patterns through logic 3. Does not perform true at-speed testing 4. Mux at chip I/O causes performance penalty 5. Routing overhead can be significant
Local boundary scan or wrapper	1. Performance penalty at the chip I/Os is avoided 2. Requires only a couple of wires to route 3. Detailed testing is possible 4. Fault diagnosis is possible	1. Test time can be large; used only for small memories 2. At-speed testing not possible 3. Wrapper causes permanent penalty in memory performance 4. Requires significant additional design effort
Built-in self-test	1. Automation tools are available 2. It simplifies ATE requirements 3. At-speed testing is possible	1. Hardware overhead is largest compared to other methods 2. Only limited number of test algorithms can be implemented 3. Fault diagnosis is difficult, area overhead with fault diagnosis is quite large 4. BIST logic causes permanent penalty in the memory performance
ASIC functional test	1. It is the easiest and least expensive method 2. There is no performance penalty 3. It does not require any design modification	1. It provides only simple functionality check 2. Fault diagnosis not possible 3. Design engineer writes this test during design verification; test engineer has no control 4. It is applicable to small memories
Through on-chip μP	1. No design modification needed and hence zero overhead 2. There is no performance penalty 3. Detailed testing is possible 4. Any memory test algorithm can be used 5. It provides at-speed testing 6. It provides failed bit location for diagnosis 7. It simplifies ATE requirements	1. It is applicable only to those SoC designs that contain one or more microprocessor cores 2. It requires an ATE interface API to handle binary code generated by the assembler

Table 7.3

Recommended Test Methodology Guidelines for a Single Embedded Memory

Memory Size	Recommended Test Method
Less than 8 Kbits	Through ASIC functional test for very small size of memory
	Direct access through mux-isolation, scan, or collar register
	Test vector generated by tester ALPG access through mux-isolation
8 Kbits to 16 Kbits	Direct access through mux-isolation and ALPG vectors
	BIST
More than 16 Kbits	BIST

Note: For multiple memories, shared controller-based BIST is recommended.

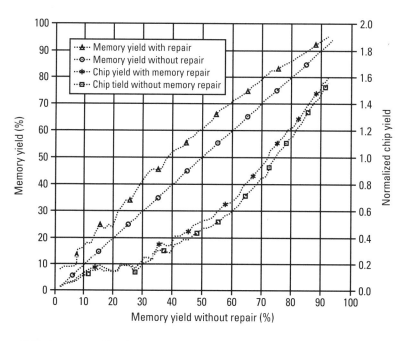

Figure 7.10 Impact on UltraSparc memory and chip yield with and without repair. (From [21], © IEEE 1997. Reproduced with permission.)

is the most widely used method for stand-alone as well as embedded memories at the present time.

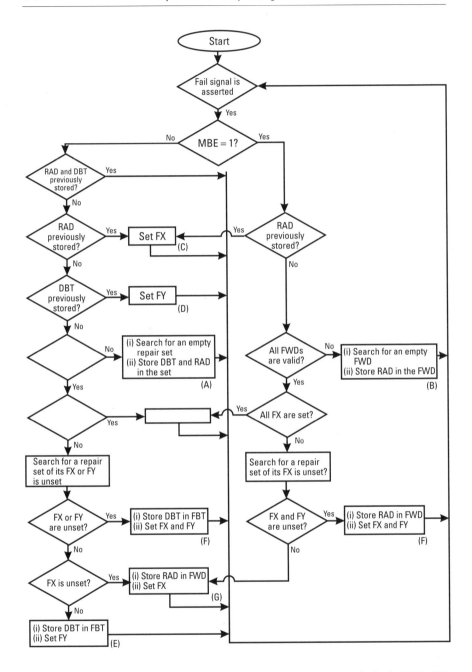

Figure 7.11 Flowchart for redundancy analysis and repair. (From [22], © IEEE 1999. Reproduced with permission.)

7.3.2 Soft Repair

Soft repair uses an address mapping procedure to bypass faulty address locations. In this scheme, BIST operation is linked with power-on of the chip—hence, every time power is switched on, memory is tested through BIST. During this testing, the addresses of all failed locations are stored separately and an address mapping procedure maps those addresses onto redundant fault-free addresses [25–27]. Although open literature recommends a dedicated finite-state machine for the implementation of address-mapping procedures, efficient implementation can also be achieved through an on-chip microprocessor or through other logic cores.

7.4 Error Detection and Correction Codes

In many SoC designs, particularly for communication networks and telephony, sophisticated error detecting and correcting codes (EDC codes) are used. These codes have been used at chip level as well as at the memory system architecture level to obtain reliable read/write operations. EDC codes can also be used to locate a faulty memory location, and spare rows/columns can be used for yield improvement as discussed in Section 7.3.

Because high-speed memories are also prone to soft errors, EDC codes are built into the chip. On-chip coding provides both benefits: protection against cell defects and protection against soft errors [28, 29]. The simplest error detection code is a parity check. For high-speed memories, a code must allow fast and parallel encoding and decoding. Due to XOR operation in the parity check circuit and penalties in speed, it has only limited use. Single-bit error correcting and double-bit error detecting codes (SEC-DED) are commonly used in high-performance SoC memories. This is because the physical design of so many memories is based upon 1-bit data input/output.

SEC-DED codes are used primarily for permanent faults. The soft error problem encountered in RAMs due to alpha particle hits is not correctable. For this purpose random double-bit error (DEC) codes have been used. The most popular codes are the basic Hamming codes using simple horizontal and vertical (H and V) parity checking and the Bose-Chaudhuri code. Bose-Chaudhuri code allows double-bit correction, however its hardware overhead is high. To avoid large area overhead, other techniques such as erase correction and read retry have also been used.

7.5 Production Testing of SoC With Large Embedded Memory

While testing for embedded memories with redundancy analysis and BISR is a necessity for yield improvement, the test implementation can have a significant impact on production costs. The primary issue is that SoC contains large memory, so should SoC be tested on a logic tester or a combination of memory and logic testers? This choice has a tremendous impact on overall SoC test costs.

Generally, the sequence of SoC testing is redundancy test, wafer sort test, pre-burn-in package test, and post-burn-in package test. In wafer sort, pre- and post-burn-in package tests, the memory and other cores/logic can either be tested on the logic tester (one pass or single insertion) or memory tester, while other cores/logic are tested on the logic tester (two passes or dual insertion). In dual insertion, the memory tester allows for parallel testing, while the number of test procedures and overhead such as lot change time and index time increases. The difference in the handler package tray for the memory tester and logic tester may also cause some logistic issues.

In [30], a cost model is reported to compare three situations:

1. Everything done on one logic tester (one tester and single insertion);
2. Redundancy test is done on the memory tester, while wafer sort and all package tests are done on the logic tester (two testers and single insertion);
3. Redundancy test is done on the memory tester. All wafer sort and package tests are done using both memory and logic testers (two testers and dual insertion).

The cost model is given by the following equation [30]:

$$T_g = [(n_l / n_s)(t + t_i) + t_{lc}]/(n_l * y) \tag{7.1}$$

where T_g is test time per good die, n_l is number of dies per lot, n_s is number of parallel sites, t is test time per die, t_i is index time, t_{lc} is lot change over time, and y is yield of wafer test. The projection by this model is shown in Figure 7.12.

The data assumed in Figure 7.12 is for a 200-pin SoC, 16 parallel sites on the memory tester, and 2 sites on the logic tester, 16-bit memory data bus, fixed die size, unity yield, 100 dice per wafer, 50 wafers per lot, 300-sec

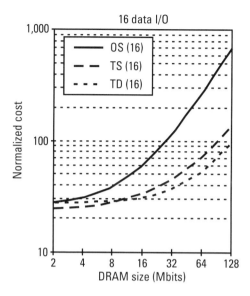

Figure 7.12 Projection of SoC test cost by Eq. (1). OS represents one tester and a single insertion; TS represents two testers and a single insertion; and TD represents two testers and dual insertion. (From [30], © SEMI 1998. Reproduced with permission.)

lot change time, 0.5-sec prober index time, 50-sec wafer loading time, lot size of packaged part is 300, memory handler index time is 16 sec, and logic handler index time is 1 sec.

Figure 7.12 illustrates a crossover in cost from single insertion to double insertion (TS versus TD). Note also that for very large memory sizes, the dual tester solution becomes much cheaper as the index time becomes relatively small compared to the actual test time. Although, the exact crossover point will vary depending on the specific parameters, this model indicates that a simple cost analysis in every individual situation is necessary and such analysis can identify a low-cost test logistic for a specific case.

References

[1]   *International Technology Road Map for Semiconductors,* Austin, TX: Sematech, Inc., 1999.

[2]   Rajsuman, R., *Digital Hardware Testing,* Norwood, MA: Artech House, 1992.

[3]   van de Goor, A. J., *Testing Semiconductor Memories,* New York: John Wiley, 1996.

[4]   Prince, B., *Semiconductor Memories,* 2nd ed., New York: John Wiley, 1991.

[5]   Sharma, A. K., *Semiconductor Memories,* New York: IEEE Press, 1997.

[6]   van de Goor, A. J., "Using March tests to test SRAMs," *IEEE Design and Test of Computers,* March 1993, pp. 8–14.

[7]   Veenstra, P. K., F. P. M. Beenker, and J. J. M. Koomen, "Testing or random access memories: theory and practice," *IEE Proc.,* Part G, Vol. 135, No. 1, Feb. 1988, pp. 24–28.

[8]   Schanstra, I., and A. J. van de Goor, "Industrial evaluation of stress combinations for March tests applied to SRAMs," *Proc. IEEE Int. Test Conf.,* 1999, pp. 983–992.

[9]   Rajsuman, R., "RAMBIST builder: A methodology for automatic built-in self-test design of embedded RAMs," *Proc. IEEE Int. Workshop on Memory Technology, Design and Testing,* 1996, pp. 50–56.

[10]   Sas, J. V., F. Catthoor, and H. J. de Man, "Test algorithms for double-buffered random access and pointer-addressed memories," *IEEE Design and Test of Computers,* June 1993, pp. 34–43.

[11]   Nadeau-Dostie, B., A. Silburt, and V. K. Agarwal, "Serial interfacing for embedded memory testing," *IEEE Design and Test of Computers,* Apr. 1990, pp. 52–63.

[12]   Hii, F., T. Powell, and D. Cline, "A built-in self-test scheme for 256Meg SDRAM," *Proc. IEEE Int. Workshop on Memory Technology, Design and Testing,* 1996, pp. 15–21.

[13]   Matsumura, T., "An efficient test method for embedded multiport RAM with BIST circuitry," *Proc. IEEE Int. Workshop on Memory Technology, Design and Testing,* 1995, pp. 62–67.

[14]   McConnell, R., U. Moller, and D. Richter, "How we test Siemens' embedded DRAM cores," *Proc. IEEE Int. Test Conf.,* 1998, pp. 1120–1125.

[15]   Ternullo, L., et al., "Deterministic self-test of a high speed embedded memory and logic processor subsystem," *Proc. IEEE Int. Test Conf.,* 1995, pp. 33–44.

[16]   Sun, X., "An integrated open CAD system for DSP design with embedded memory," *Proc. IEEE Int. Workshop on Memory Technology, Design and Testing,* 1998, pp. 4–9.

[17]   Rajsuman, R., "A new test methodology for testing embedded memories in core based system-on-a-chip ICs," *Proc. IEEE Int. Workshop on Testing Embedded Cores Based System,* 1998, pp. 3.4.1–3.4.6.

[18]   Bhavsar, D. K., and J. H. Edmondson, "Testability strategy of the alpha AXP 21164 microprocessor," *Proc. IEEE Int. Test Conf.,* 1994, pp. 50–59.

[19]   Dreibelbis, J., et al., "Processor based built in self test for embedded DRAM," *IEEE J. Solid State Circuits,* Vol. 33, No. 11, Nov. 1998, pp. 1731–1740.

[20] Saxena, J., et al., "Test strategy for TI's TMS320AV7100 device," *Proc. IEEE Int. Workshop on Testing Embedded Cores based Systems,* 1998, pp. 3.2.1–3.2.6.

[21] Youngs, L., and S. Paramanandam, "Mapping and repairing embedded memory defects," *IEEE Design and Test,* Jan.–Mar. 1997, pp. 18–24.

[22] Nakahara, S., "Built-in self-test for GHz embedded SRAMs using flexible pattern generator and new repair algorithm," *Proc. IEEE Int. Test Conf.,* 1999, pp. 301–310.

[23] Niggemeyer, D., J. Otterstedt, and M. Redekker, "A defect tolerant DRAM employing a hierarchical redundancy, built-in self-test and self-reconfiguration," *Proc. IEEE Int. Workshop on Memory Technology, Design and Testing,* 1997, pp. 33–40.

[24] Kim, I., et al., "Built in self repair for embedded high density SRAM," *Proc. IEEE Int. Test Conf.,* 1998, pp. 1112–1119.

[25] Chen, T., and G. Sunada, "Design of a self-testing and self-repairing structure for highly hierarchical ultra large capacity memory chips," *IEEE Trans. VLSI,* Vol. 1, No. 2, June 1993, pp. 88–97.

[26] Bair, O. S., et al., "Method and apparatus for configurable built-in self-repairing of ASIC memories design," U.S. Patent No. 5577050, Dec. 28, 1994.

[27] Kablanian, A., et al., "Built-in self repair system for embedded memories," U.S. Patent No. 5764878, Feb. 7, 1996.

[28] Fujiwara, E., and D. K. Pradhan, "Error control coding for computers," *IEEE Trans. Computers,* July 1990, pp. 63–72.

[39] Rao, T. R. N., and E. Fujiwara, *Error Control Coding for Computer Systems,* Englewood Cliffs, NJ: Prentice Hall, 1989.

[30] Shitara, T., "Strategies for minimizing cost for testing system-on-a-chip," *Proc. Semicon West Manufacturing Test Conf.,* SEMI, 1998, pp. B.1–B.6.

8

Testing of Analog and Mixed-Signal Cores

Testing of embedded analog circuits in SoC is a significant challenge. The lack of controllability/observability is the primary difficulty. Each embedded analog/mixed-signal component has an extremely limited number of pins (in most cases fewer than five, in some cases just one or two). The direct test application and response evaluation is thus almost never possible. Also in SoC designs, the interaction between the digital and analog portions is much closer, so it is often necessary to access the analog portion through complex digital interfaces. A good example of this situation is Analog Devices' AD7015, which is used in digital cellular telephones. It has 11 digital-to-analog converters (DACs), plus filters and logic on a single chip. In the transmit section of this device, 10-bit DACs are surrounded by other functional blocks and, thus, a conventional linearity test cannot be applied without special provision.

Another major difficulty that arises when testing analog circuits is that analog faults do not cause a simple state change in the logic values. Some common analog test issues are as follows:

- Analog signals are tested within specified bands/limits.
- Analog signals are sensitive to process variations and are thus performance-sensitive to the process (parameter variation, correlation, mismatch, noise, and so on). This prevents the abstraction of a standard analog fault model if the manufacturing process changes.

- Lack of analog design-for-test (DFT) methodologies.
- Specification-driven testing of analog circuits, manual test generation, and lack of robust EDA tools, resulting in long test development times.
- The yield versus defect level trade-off in analog testing is nondeterministic, because most analog faults do not result in catastrophic failure.
- Analog test results are affected by noise and measurement accuracy. Also, pass/fail results do not provide any diagnosis capability and are inadequate, preventing the use of analog BIST methods. This item is further complicated because any intrusion of a DFT circuit into an analog circuit affects the circuit's performance. Thus, it is difficult to develop analog BIST and DFT methods.

In this chapter, we describe the testing of some analog cores commonly used in SoC designs. In Section 8.1, we describe testing parameters and characterization methods for these circuits. Design-for-test and built-in self-test methods for these cores are discussed in Section 8.2.

8.1   Analog Parameters and Characterization

The most commonly used analog cores in SoC designs are DAC, ADC, and PLL. Each of these circuits has a separate set of design parameters and hence different testing methods. Definitions of design parameters were given in Chapter 3 for various analog circuits. In this section, we describe characterization and testing methods for these parameters.

8.1.1   Digital-to-Analog Converter

Digital-to-analog converters (DACs) are commonly used to generate output from the SoC—such as audio signals or video data—to the outside world (for example, triplet DACs for RGB colors). These DACs are designed independently from the digital logic, are fully characterized on a fixed fabrication process, and are used as hard cores. Various DAC parameters are tested as follows [1]:

Offset Error

The offset error (V_{OS}) of a DAC is measured as the output voltage when a digital input corresponding to the null code is applied. For a high-precision DAC, it is a very small voltage and requires a high accuracy digital voltmeter.

An alternate measurement could be made by first amplifying the output or by using a nulling DAC.

Full-Scale Range

The FSR is calculated from the offset and full-scale measurements as:

$$FSR = V_{FS} - V_{OS}$$

Gain Error

The full-scale *Value* (V_{FS}) at the DAC output is measured by applying the full-scale digital code at its input. The gain error is then computed as:

$$\text{Gain error} = (V_{FS} - V_{OS}) - V_{FSR(ideal)}$$

LSB Size

The LSB size is calculated from the offset and full-scale measurements as:

$$LSB = FSR/(2^N - 1)$$

where N = number of bits (resolution).

DNL Error

The LSB step between any two adjacent codes is measured by the difference in output voltage obtained by applying the first code and allowing the DAC to settle before applying the next code, then letting it settle to the new voltage level. The DNL (for these adjacent codes) is calculated as the difference between this measured LSB step and the ideal LSB size. In production testing, it may take too long to test DNL at each code, and so few selected codes are used based on the DAC architecture (such as major transitions for an $R/2R$ ladder, all codes for a video DAC, and so on).

INL Error

INL can be calculated from the DNL code measurements as the code value that is farthest from the straight line (end point, or best fit). The end-point line is given as:

$$y = mx + c$$

where m is LSB size or

$$m = FSR/(2^N - 1) \text{ and } c = V_{OS}$$

Reference Voltage

If the DAC has an internal reference voltage that is brought to an external pin, it is directly measured. If it is not brought out, it is calculated indirectly from the full-scale range measurement. If the DAC requires an external reference, a high accuracy and low drift reference must be supplied during testing.

SNR, SINAD, and THD

Testing the dynamic parameters of a DAC requires digital input codes to be applied that correspond to a sine wave, coherent sampling of the DAC output, and DSP frequency analysis to compute the signal, harmonics, and noise power. Coherent sampling (for both the input digital codes and output samples) must satisfy the following condition:

$$F_s / N = F_t / M$$

where F_s is the sample frequency, F_t is the test frequency, N is the number of samples, and M is the number of cycles. The output of the DAC should be smoothed with a filter to remove the steps from the sine wave before taking the samples. For measuring high-resolution DACs, the output may need to be further conditioned by filtering out the fundamental frequency with a notch filter and amplifying the remaining signal (noise and harmonics). The primary reason is that the frequency response of DACs may suppress noise or harmonics that need to be measured. The SNR, SINAD, and THD can be calculated from the fast Fourier transform (FFT) spectrum.

Intermodulation Distortion

Intermodulation distortion (IM) requires digital code inputs of a multitone (usually two-tone) sine wave, instead of a single-frequency sine wave. The rest of the test conditions are the same as those mentioned above.

The characterization of a DAC depends upon the DAC's architecture, application, resolution, and speed. Also, all of the characterization parameters are not measured during the production testing. An example of characterization and production test parameters is given in Table 8.1. As listed in Table 8.1, only a small number of parameters are tested during production.

8.1.2   Analog-to-Digital Converter

Analog-to-digital converters (ADCs) are commonly used in SoC to receive input such as microwave or radio signals from the outside world. Similar to DACs, ADCs are designed separately from the digital logic, are fully

Table 8.1

Commonly Used DAC Parameters for Characterization and Production Testing

Parameters	Characterization	Production Test
DC	Resolution	Resolution
	Integral linearity error and drift	Integral linearity error and drift
	Differential linearity error and drift	Differential linearity error and drift
	Gain error and drift	Gain error and drift
	Unipolar and bipolar errors	—
	Grayscale error	—
	Monotonicity	—
	LSB size	LSB size
	Lout, Rout, and Cout	—
	Output compliance and noise	Output compliance and noise
Logic	VIH, VIL, IIL, IIH	VIH, VIL, IIL, IIH
Power supply	Vcc, Vdd	Vcc, Vdd
	Supply current	Supply current
	Power dissipation	Power dissipation
	Power supply rejection ratio (PSRR)	PSRR
	Signal-to-noise ratio (SNR)	SNR
	THD	
AC	Clock rate	Clock rate
	Settling time	—
	Delay, rise time, fall time	—
	Glitch impulse	Glitch impulse
	Differential gain and phase error	Differential gain and phase error
Reference	Reference output voltage	Reference output voltage
	Maximum reference output load	Max reference output load

characterized in the fabrication process, and are used as hard cores. Various ADC parameters are tested as follows [1–5]:

Offset Error

Zero-scale value is obtained by measuring the transition voltage from the zero (null) value to the first output code (V_0), and taking its difference with voltage for ½ LSB size (which is obtained after the full-scale measurement). The offset error is its difference from the ideal zero-scale value ($\frac{1}{2}$ LSB$_{ideal}$).

Full-Scale Voltage Error

The full-scale voltage is obtained by measuring the transition voltage of the last binary bit transition to all ones output ($V(2^N - 1)$) and adding ½ LSB. Its difference from the ideal full-scale voltage gives the error.

LSB Size

LSB size is calculated as $(V(2^N - 1) - V_0)/(2^N - 2)$.

Gain Error

Gain error is calculated as full-scale voltage error minus the offset error.

Static DNL and INL

These values can be computed from the transition points for successive codes. For DNL, the code width is computed as the difference of the two successive code transitions. DNL is computed as the maximum of the difference between any code width and the device LSB size. For INL, the center points of each code (midpoint of the two adjacent transitions) are compared with the straight line (end point or best fit). Testing all codes may be too time consuming, but some ADCs can be tested for linearity by testing only major transitions and any decoded upper bits.

Dynamic DNL and INL

A common method to test dynamic DNL and INL is the code-density test, also known as the histogram test. It involves collecting a statistically significant number of digitized samples over a period of time in order to build a model of the ADC's response to a specific well-defined input (usually a ramp or a sine wave). The difference in the probability density function (pdf) of the ADC to the input signal is used to compute DNL and INL. The pdf of a ramp is a uniform function and the pdf of a sine wave of amplitude A, is given by

$$\frac{1}{\pi\sqrt{\left(A^2 - V^2\right)}}$$

DNL is calculated as:

$$\frac{\text{Actual } P(n\text{th code})}{\text{Ideal } P(n\text{th code})} - 1$$

INL can be computed after obtaining the DNL for all codes. Any empty histogram bin indicates a missing code. For both ramp and sine inputs, the amplitude is chosen that slightly overdrives the ADC. Coherent sampling must be used.

SNR, SINAD, THD, and IM

These parameters can be measured using an FFT-based test that requires the application of a sine-wave input (two-tone input for IM), and the performance of a DSP frequency analysis on the ADC output to compute the signal, harmonics, and noise power. The input sine wave must be selected carefully for each DUT because it must be offset by the same amount as the DUT offset and have an amplitude as large as possible without clipping. The frequency must be timed coherently with the sample frequency. A track and hold is needed at the input of the ADC if it does not have an internal one. The various parameters can be computed from the FFT spectrum similar to the method used for DAC.

Effective Number of Bits (ENOB)

The ENOB can be calculated from the SNR as follows: ENOB = (SNR − 1.76)/6.02. Also, a sine-wave curve-fitting algorithm can be used in accordance with IEEE Standard 1057. If the frequency of the input sine wave is known, the algorithm is used to determine the amplitude, phase, and offset of the best-fit sine wave, which is used as a reference from which to calculate the rms error if the measured data can be calculated. ENOB is given by $N - \log_2$ (measured rms error /ideal rms error).

Aperture Errors

Aperture errors include jitter, uncertainty, and delay. The locked histogram test can be used to measure the aperture jitter and uncertainty of an ADC. The same frequency is used for the input sine wave and the sampling clock, which are synchronized such that the same point on the input is sampled repetitively (usually midscale, where the rate of change is maximum) and a histogram of the output codes is obtained. The standard deviation, σ, of the histogram is used to calculate the aperture error as follows:

$$T_a = \sigma(V_{\text{LSB}}/\text{slew rate})$$

where

$$\text{slew rate} = 2\pi V_{amp} f_{in}$$

The locked sine-wave test can also be used to measure the effective aperture delay by measuring the difference between the leading edge of the sampling clock pulse and the actual zero crossing of the sine wave input.

ADC testing requires more statistical work than DAC testing. Testing an ADC by applying a DC analog input voltage or step and measuring the output digital code is a very poor test. In measuring the static parameters of an ADC, the precise transition points where the code changes occur must be measured. This is generally done by using a servo loop method.

In the servo loop test, a closed-loop system is used that seeks to locate the statistical code edges of each code at a time. This is done by placing a digital code in the register of the servo loop system so that the closed loop determines the transition voltage as shown in Figure 8.1(a). Because this is a very slow test, in practice ATEs use faster variations that are based on the same closed-loop principle, but use a digital feedback loop as shown in Figure 8.1(b). All test parameters are calculated from measured transition points.

8.1.3 Phase-Locked Loop

Testing of a phase-locked loop (PLL) varies depending on whether it is fully integrated or uses an external loopback filter. It is also dependent on whether any design-for-test techniques have been used that facilitate access to internal

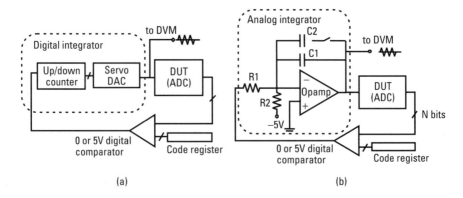

Figure 8.1 (a) Analog servo loop test for ADC. (b) Digital servo loop [5].

nodes, and whether the PLL can be opened during test mode [6, 7]. Another factor that affects testing is the ATE functionality (mixed-signal functions) and the maximum frequency of the tester compared to the PLL frequency range. Various PLL parameters are tested as follows:

Settling Characteristic (Phase Step Response)

This is tested by measuring the output response to a step in phase of the input signal. The phase of the input reference clock is shifted by π. The input and output of the PLL are captured simultaneously and the phase error is calculated.

Closed-Loop Gain/Phase Transfer Function

The settling data in the time domain from the phase step response is used to evaluate the closed-loop frequency characteristics via frequency conversion (FFT). The time-domain data is captured with high accuracy.

Lock range

First the PLL is set to its locked condition at the VCO center frequency and the input frequency is slowly increased until the PLL cannot follow it and becomes unlocked. This gives the upper frequency limit of the lock range. This procedure is repeated with decreasing frequencies to get the lower limit. This measurement is repeated for each divide number and appropriate phase-detector gain.

Capture Range

The opposite of the lock range test is done to measure the capture range. That is, the test starts with the PLL in the unlocked condition and the frequency is slowly changed until it locks, both in the upper and lower frequency ranges.

VCO Transfer Function

This can be measured only if test access is available to the VCO input and the external loop can be disconnected. A ramp input is applied as a control voltage to the VCO and its output frequency is measured to obtain the VCO frequency versus voltage transfer function.

VCO Center Frequency

To measure this frequency, the two pins of the symmetrical VCO input are grounded and the VCO oscillating frequency is measured.

VCO Gain

For VCO gain, one of the symmetrical VCO inputs is grounded and a variable DC voltage is applied to the other. This signal corresponds to the loop filter output signal (U_f). The VCO gain is equal to the variation of the VCO angular frequency related to the variation of the U_f signal by $\Delta U_f = 1$V.

Natural Frequency and Damping Factor (ω_N and ζ)

To measure the natural frequency and damping factor, a disturbance is applied to the PLL. It is done by modulating the reference frequency with a square-wave signal. The transient response of the PLL is analyzed by the loop filter output signal through which ω_N and ζ are computed. These values can also be used to compute the lock range ($\Delta\omega_L = 2\zeta\omega_N$) and VCO gain to validate the measured values.

Phase Detector Gain (K_d)

For phase detector gain, the VCO output signal is coupled capacitively to one of the phase detector inputs and a variable DC level is applied to the other input. When the level of applied DC voltage is equal to the average value of previously applied sine signal, the output signal of the phase detector becomes a square wave having peak amplitude K_d.

Another method for this measurement is to inject current into the phase detector output and measure the phase shift between reference and VCO output while the PLL is locked.

Phase Transfer Function and 3-dB Bandwidth

The phase transfer function is obtained by measuring the amplitude of the loop filter output signal at different modulation frequencies. A sweep generator generates the modulation frequency and a spectrum analyzer can do the measurement. The 3-dB frequency is determined from the transfer function.

Pull-In Range

For this, the reference frequency is set to approximately center frequency and then slowly increased until the loop locks out; the frequency is then slowly decreased to monitor when the loop pulls in again.

Jitter

The most commonly specified jitter parameter is cycle-to-cycle jitter. Unfortunately, it is difficult to measure directly with most measurement equipment for high-performance PLLs. As a result, different methods are used that approximate cycle-to-cycle jitter.

The typical method of measuring the jitter is to accumulate a large number of cycles, create a histogram of the edge placements, and record peak-to-peak as well as standard deviations of the jitter. Care must be taken that the measured edge is the edge immediately following the trigger edge. The oscilloscope cannot collect adjacent pulses; rather it collects pulses from a very large sample of pulses. This method produces period jitter values somewhat larger than if consecutive cycles (cycle-to-cycle jitter) were measured.

Dynamic Idd

While PLL is in the locked state, its input is turned off. The dynamic Idd is then measured at minimum VCO frequency (PLL is free running and not locked).

8.2 Design-for-Test and Built-in Self-Test Methods for Analog Cores

Analog design for testability methodologies can be classified into two types: accessibility methods and reconfiguration methods.

Accessibility DFT methods rely on a test bus and/or test point insertion for controllability and observability of the internal nodes of the circuit under test. This constitutes one of the earliest approaches to the mixed-signal test problem with isolated and ad hoc solutions on a need basis [8]. The proposed IEEE 1149.4 mixed-signal test bus could also be viewed as an accessibility method. Recently, Opmaxx (now Fluence Technology) has commercialized a test point insertion scheme (licensed from Tektronix) that inserts analog probes to obtain access to internal points. Reconfiguration DFT methods rely on the reconfiguration of the circuit under test for test-ability improvement. The impact of these methods on circuit performance is generally high. These methods are still in the research phase and have rarely been used in mass production.

Analog BIST methods can also be classified into two categories: func-tional methods and fault-based methods. The functional methods use tradi-tional stimuli (e.g., sinusoidal or multitone waveforms) and measure the functional specifications of the circuit. Fault-based methods target the detec-tion of specified faults and use nonconventional stimuli and signatures. Most of the BIST methods test the analog circuits in a mixed-signal environment, such as via a digital–analog–digital path, unlike the analog–digital–analog path used by conventional external tests.

In the last few years, a number of research papers have appeared that propose analog BIST methods [9–12]. Recently Fluence Technology and LogicVision have also announced analog/mixed-signal BIST solutions; these are described in the following subsections.

8.2.1 Fluence Technology's Analog BIST

Fluence's BISTMaxx is based on Opmaxx's oscillation BIST technology. It is a vectorless BIST scheme that obviates the need for test stimuli selection and application. It converts the circuit into an oscillation mode during test as shown in Figure 8.2. Faults cause the circuit to either stop oscillating or the oscillation frequency to differ beyond the tolerance range of its nominal value. The oscillation frequency can be evaluated using pure digital circuitry and interfaced to the boundary scan [13].

BISTMaxx can be used to test a variety of analog mixed-signal circuits ranging from converters and oscillators to PLLs and amplifiers. The test method consists of partitioning the complex circuit into functional building blocks such as operational amplifier, comparator, filter, and so on, which are then converted into circuits that oscillate by adding some additional circuitry. Because BISTMaxx is digitally invoked, it is a relatively nonintrusive solution that performs structural and functional tests.

Two additional capabilities (licensed from Tektronix) are recently combined with BISTMaxx. These are a histogram-based BIST method for testing ADCs and analog probing that provides test points inside the mixed-signal device with external access through a multiplexing scheme.

8.2.2 LogicVision's Analog BIST

LogicVision has commercialized ADC BIST as well as PLL BIST [14].

8.2.2.1 adcBIST

The adcBIST product is based on an all-digital BIST method used for testing flash or SAR ADCs. Connections are made to only the inputs and outputs of the ADC to be tested. An analog multiplexer is typically inserted in the input path. The operation is kept fully synchronous to the sampling clock. The other analog components used are internal/external R and C (20% accuracy) to convert the digital stimulus (a ramp) from the BIST controller to an analog input. The basic structure is shown in Figure 8.3.

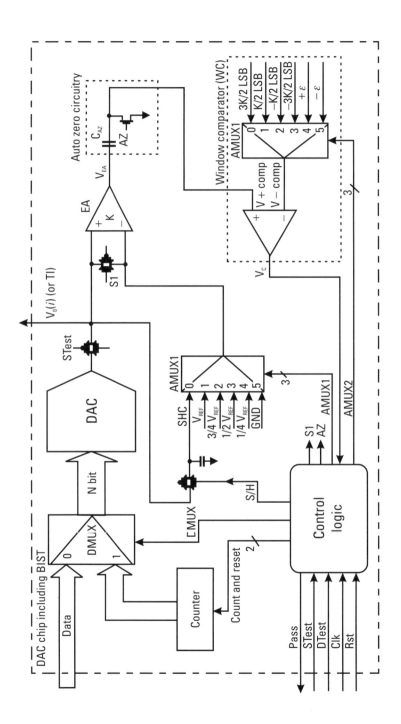

Figure 8.2 Basic structure of Fluence's BISTMaxx and how it is used. (From [10], © IEEE 1998. Reproduced with permission.)

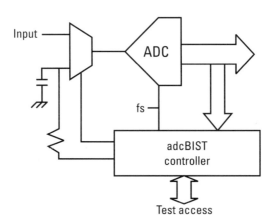

Figure 8.3 Basic structure of LogicVision's adcBIST [14]. In the normal mode, input is a sine wave; in the test mode it is a ramp signal.

The output of the ADC, as it samples the stimulus, is processed by the adcBIST circuitry that produces six intermediate results from which the offset, gain, second- and third-harmonic distortion, and idle channel noise are computed. Depending on the area overhead requirements, three options are provided. In the minimum configuration, the intermediate results are output to an 1149.1 TAP controller or other test access circuit. In this mode, an ATE performs arithmetic calculations to determine the ADC test parameters. In the standard configuration, the controller itself computes the ADC parameters; and in the maximum configuration, the comparison to limits is also done on-chip. A software tool generates Verilog/VHDL for adcBIST controller; the size of controller is in the range of 900 to 1400 gates for a 12-bit ADC. The main limitation of adcBIST is that it does not test INL and DNL, which are very important parameters.

8.2.2.2   pllBIST

LogicVision's pllBIST has modes for testing PLLs containing level-sensitive phase comparators (e.g., exclusive-or) and edge-sensitive phase-frequency comparators (e.g., charge-pump). It uses the externally supplied reference clock input to the PLL as a constant frequency stimulus and also generates an on-chip stimulus for connection to the PLL input via an on-chip multiplexer. A constant rate pseudorandom bit stream is supplied instead of a clock—however, an alternate reference clock is needed (any constant frequency). Clock outputs of the PLL and a lock indicator are processed by pllBIST. Digital results for the loop gain and upper and lower lock range frequencies are output to an 1149.1 TAP. The structure of pllBIST is shown in Figure 8.4.

An additional option measures the peak-to-peak and/or rms jitter. The cumulative distribution function (CDF) limits for the jitter (0% to 100% for peak-to-peak; 16% to 84% for rms) are loaded in at test time from off chip or are hard-wired onto the chip. By loading in various limits, the entire CDF can be measured. Capture/lock time is also measured. The area overhead of pllBIST controller is in the range of 1000 to 1500 gates. The main limitation of this method is the fact that frequency jitter and phase errors are not tested.

8.2.3 Testing by On-Chip Microprocessor

If we consider the usage of ADC/DAC in embedded context as cores, ADC inputs are generally accessible by the primary I/Os but its outputs are not. On the other hand, DAC outputs are generally accessible at the primary I/O level but its inputs are not. Thus, the simplest form of testing can be done as follows:

1. For DAC, inputs are not accessible, hence test stimuli can be generated on chip, while performing response evaluation off chip because outputs are accessible. Thus, the test stimulus should be generated by the on-chip microprocessor for embedded DAC, while response can be evaluated by the ATE.

2. For ADC, inputs are accessible, hence, test stimuli are applied from ATE, while performing test response evaluation on chip because outputs are not accessible.

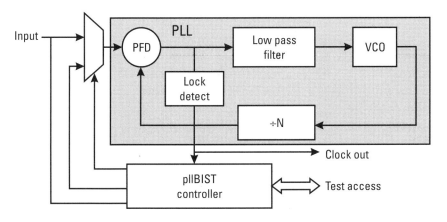

Figure 8.4 Basic structure of LogicVision's pllBIST [14].

Under the assumption that the on-chip microprocessor is fault free, the testing of ADC/DAC can be done by the microprocessor. For this testing purpose, an assembly language program (written in the microprocessor's assembly language) is developed that generates the necessary test stimuli when executed by the microprocessor (we will call it the test generation part). Such a program is converted into binary code using a microprocessor assembler. The binary code is stored in ATE memory and with an interface circuit it is applied to the microprocessor. This method is described in [15].

Because the binary code is microprocessor instructions and data, the microprocessor executes them and generates the desired test stimuli for the ADC/DAC. The response of the ADC/DAC against this test stimulus is either evaluated concurrently by the microprocessor or stored in the on-chip memory for later evaluation or done by the ATE. In the case when evaluation is also done by the microprocessor, it is done by executing another program on the microprocessor that is similarly developed in assembly language, converted into binary, and applied to the microprocessor (we will call it the test-evaluation part). In response to this evaluation program, the microprocessor performs the necessary calculations to evaluate the ADC/DAC response and determines if a parameter is outside its specification (presence of analog fault).

Note that if the on-chip memory is not sufficient to store the ADC/DAC response, the response could be stored in the ATE memory. In such a case, the test response evaluation can be done by the ATE to determine pass/fail.

To apply the test data generated by the microprocessor core to ADC/DAC, one register is required. In the test mode, the contents of this register can be altered by index addressing, such as by addressing through any one of the microprocessor address registers. The implementation of this register can be (1) an independent register or (2) a dedicated memory location. In the test mode, this register (memory location) provides the inputs to the ADC/DAC via a multiplexer. During normal mode this register is cut off via the same multiplexer. Note that the test register and multiplexer can be either implemented independently or within the on-chip bus, if such a bus exists on the chip. The concept is illustrated in Figure 8.5.

The overall sequence of operation at SoC level is given as follows:

1. Test the on-chip microprocessor and memory.

2. Develop an assembly language program that can generate the desired test input for DAC/ADC.

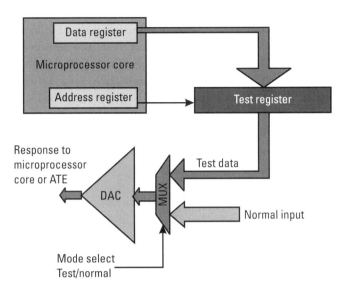

Figure 8.5 Mechanism for test data application from the microprocessor core to DAC.

3. Use the assembler of the microprocessor core to generate the object code of the assembly language program.

4. Apply the object code to the microprocessor core through an interface circuit.

5. The microprocessor core generates the DAC/ADC test patterns and applies them to the DAC/ADC.

6. Collect and evaluate the response: (a) Response is collected in on-chip memory. The microprocessor core executes the test-evaluation part of the program and computes specific parameter values. Based on these values, the microprocessor determines pass/fail and sends this information to the host computer or tester. (b) The response is collected in the memory of the host computer or tester. The host computer or tester executes a program to compute various parameters and determines pass/fail. This program is similar in nature to the evaluation part; however, it need not to be in the assembly language.

8.2.4 IEEE P1149.4

The proposed IEEE P1149.4 Standard for a Mixed-Signal Test Bus is an extension of the 1149.1 Boundary Scan (JTAG) standard that is aimed at standardizing the architecture for and the method of access to the analog portion of mixed-signal circuits and boards for test and diagnostic applications [16–19].

The current P1149.4 proposal uses a six-wire bus, with four of the signals being the same as those used by the 1149.1 test access port (TAP). The 1149.1 bus carries the control information for setting up the test architecture to perform a measurement. The two additional wires, AT1 and AT2, form an analog stimulus-and-response bus called the *analog TAP* (ATAP). These signals can be connected to any functional pin inside the device, through an analog boundary module (ABM) via internal analog buses, AB1 and AB2. AT1 and AT2 are bused at the system level and would be connected to the external ATE. The structure is shown in Figure 8.6.

The ATAP consists of the AT1 and AT2 pins. AT1 must support a current stimulus and AT2 must support monitoring a voltage. Optionally, two more analog bus pins may be added to support differential measurement. The ATn bus is connected to the internal ABn bus through a special collection of switches and control logic called the *test bus interface circuit* (TBIC). The ATAP pins can emulate the behavior of 1149.1 pins, which allows the ATE to validate their integrity before using them for testing. Figure 8.7 illustrates TBIC and ABM.

The analog boundary module (ABM) is a collection of conceptual analog switches and digital controls that allow for the routing of test signals to

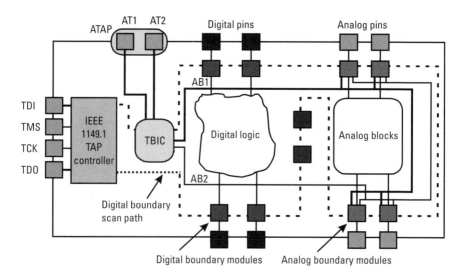

Figure 8.6 Basic structure of P1149.4.

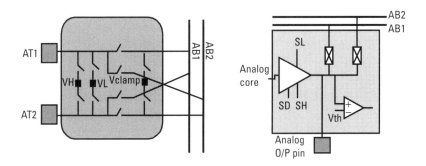

Figure 8.7 Test bus interface circuit and analog boundary module. (From [16], © IEEE 1997. Reproduced with permission.)

and from any functional pin. A scannable boundary data register provides the control signals SH, SL (drive a high or low or provide a current path for interconnect test), SD (core disconnect or high-impedance state), SB1, and SB2 (connecting ATn to ABn). A comparator digitizes the value at the pin, which can be captured in the data register. Transmission gates or tri-stateable buffers may implement the switches. Proposed standard 1149.4 contains PROBE and enhanced EXTEST instructions with MEASURE and INTEST instructions. The PROBE instruction is similar to the SAMPLE instruction, but it allows mission-mode monitoring of a signal over continuous time, providing a real-time virtual scope probe facility. During EXTEST, the ABn internal bus can be connected to the ATn external bus via TBIC control to perform a MEASURE. The 1149.4 INTEST instruction allows the analog pins to be connected to the external AT bus via the ABMs and AB bus for probing (one or two pins at a time).

The 1149.4 test bus is designed to measure board-level impedance from 10 to 100 kΩ with an accuracy of 1%. The accuracy degrades outside this impedance range. Hence, if it is used in the SoC, a modification is required to compensate for change in impedance. Another limitation is the relatively low bandwidth of 100 kHz supported by the transmission gate-based analog bus. There are various sources of DC and AC measurement errors caused by finite and nonzero switch impedance, nonlinearity, and noise. These errors can be minimized by certain design practices, including items such as using current/voltage buffers instead of CMOS transmission gates for switching AT1/2 to functional pins, using differential signaling to reduce the impact of noise, and layout of an AC ground pin between AT1 and AT2 to reduce capacitive coupling. Due to the various loading effects of AT1 and AT2 that may cause measurement errors larger than 30% or greatly reduce the bandwidth below 10 kHz, all measurements made on the 1149.4 bus must be made relative to calibration measurements.

8.3 Testing of Specific Analog Circuits

8.3.1 Rambus ASIC Cell

The most challenging aspect of testing the Rambus ASIC cell (RAC) is its high-speed interface, which requires special fixturing and timing accuracy to obtain precise measurements. The test method is described in [20].

The tester channels can be connected to the DUT using a dual transmission line (DTL) arrangement with the tester driving into a 50Ω line that is parallel terminated at the receiver. The DUT drives two parallel 50Ω terminated lines. To facilitate fast I/O switching, the bus is operated in an open drain fashion (normally driven high by the tester, pulled low by the DUT when needed), which avoids driver on/off delays and allows testing the zero bus cycle turn-around requirement for the DUT. A special probe setup using a 50Ω microstrip probe card, high-speed probes, and a high-bandwidth sampling oscilloscope was used in [20]. Care is also needed to achieve automatic time-domain reflectometer (TDR) calibration of the DUT fixture to an accuracy of ±80 ps for both drivers and receivers. All the registers in the RAC cell are scannable and a bypass mode is available for the PLL.

RAC uses on-chip PLLs (or DLLs) for cancellation of I/O delays. For production testing, the worst case jitter value for the PLL may be obtained by integrating its output waveform with a digital storage scope while running some test patterns. For characterizing the PLL phase response to various noise sources, a detector consisting of an RF mixer, a low-pass filter, and a scope is used. The RF input of the mixer is driven by the PLL output via the probe, and the local oscillator (LO) input of the mixer is driven by a reference clock from the tester. The output of the mixer (XOR of the two inputs) is low-pass filtered to obtain a real-time signal proportional to the PLL phase. Because the PLL portion is designed using a differential source-coupled-logic design style, it requires on-chip current measurement in addition to voltages. Instead of using invasive current measurement, a single force probe can be used to plot the I-V characteristics using the tester PMU. The output impedance and short-circuit current can be determined from this measurement.

In addition to the measurement of the timing and voltage of the RAC interface signals, the pin capacitance is also an important parameter to be measured. Because the typical input capacitance is on the order of 1.8 pF, which is too small to be measured accurately with most instruments, the

TDR is used to measure the reflected waveform from the DUT pin. The capacitance is proportional to the area of the TDR waveform. This can achieve accuracy on the order of 0.2 pf.

8.3.2 Testing of 1394 Serial Bus/Firewire

The high-speed serial physical interface (PHY) represents the most difficult part of the Firewire test. The challenge is due to the high data rates and the low differential transmit/receive levels. Another complication is due to the common mode signaling for speed signaling and port status detection. The test solution must have the ability to generate high-speed, low-level differential drive signals and compare small-amplitude signals of approximately 200 mV. This must be accomplished with the required precise edge placement and compare strobe timing accuracy and must also allow for at-speed generation of the additional drive levels needed for speed signaling.

An ideal test environment would require differential current drivers, differential receivers, two programmable current sources to emulate common mode speed signaling, a programmable reference voltage for TpBias, single ended drive/compare capability, and a pair of 55Ω transmission lines with 55Ω drivers and termination as shown in Figure 8.8. The test patterns for this ideal environment are shown in Figure 8.9. However most ATE provides transmission line connections with impedance of 50Ω and most drivers and comparators are single ended and not differential.

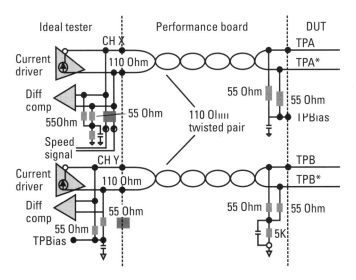

Figure 8.8 Ideal solution for Firewire interface testing.

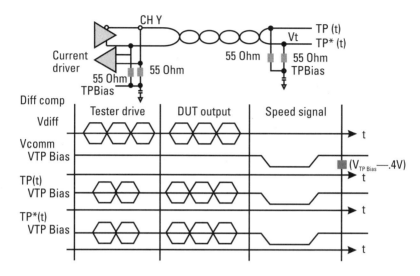

Figure 8.9 Ideal test patterns for Firewire interface testing.

A test solution is described in [21] that uses standard ATE interfaces to test Firewire at speed. Single-ended tester channels are used in a fly-by configuration to achieve the differential test capability as shown in Figure 8.10. The test patterns for this case are given in Fig. 8.11. Two tester channels are

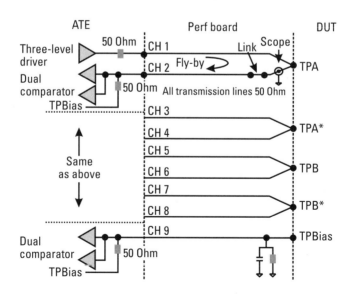

Figure 8.10 ATE solution for Firewire interface testing using fly-by architecture.

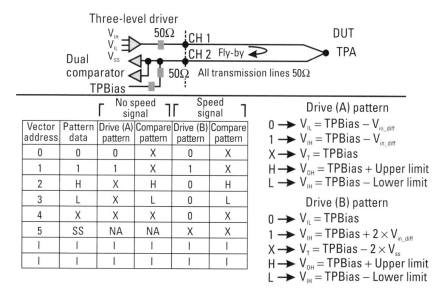

		No speed signal		Speed signal	
Vector address	Pattern data	Drive (A) pattern	Compare pattern	Drive (B) pattern	Compare pattern
0	0	0	X	0	X
1	1	1	X	1	X
2	H	X	H	0	H
3	L	X	L	0	L
4	X	X	X	0	X
5	SS	NA	NA	X	X

Drive (A) pattern

$0 \rightarrow V_{IL} = TPBias - V_{in\_diff}$
$1 \rightarrow V_{IH} = TPBias - V_{in\_diff}$
$X \rightarrow V_T = TPBias$
$H \rightarrow V_{OH} = TPBias + Upper\ limit$
$L \rightarrow V_{IH} = TPBias - Lower\ limit$

Drive (B) pattern

$0 \rightarrow V_{IL} = TPBias$
$1 \rightarrow V_{IH} = TPBias + 2 \times V_{in\_diff}$
$X \rightarrow V_T = TPBias - 2 \times V_{ss}$
$H \rightarrow V_{OH} = TPBias + Upper\ limit$
$L \rightarrow V_{IH} = TPBias - Lower\ limit$

Figure 8.11 The 400-MHz ATE test patterns for Firewire interface testing.

connected to each DUT pin (TPA/A∗, TPB/B∗), one for the drive channel and the other for comparison. Two drive channels and two compare channels provide the complementary signals to the differential pair. The drive/compare data on TPA (TPB) are always the exact complement of TPA∗ (TPB∗). The Firewire specification requires a 55Ω transmission line terminated in a 55Ω load. With the fly-by arrangement, however, the DUT pin is connected to two 50Ω transmission lines, one of them replacing the load. This provides a perfect terminated transmission line environment without adding any performance board circuitry. This does result in the need to scale the voltage derived from the DUT output currents by approximately 10%.

A tri-level driver is used that can drive Vil and Vih during input periods and drive a Vt level (say, TpBias) when the DUT output is expected. The DUT output is thus superimposed on a consistent level and can be sampled with a fixed threshold.

Another test that needs to be performed is for speed signaling. Generating small drive levels and being able to detect levels of 200-mV peak-to-peak is not easily accomplished by a general-purpose ATE. It is achieved with the fly-by configuration such that the channel used to capture the DUT output waveform can also be used to look at the tester drive waveforms. The TPA pair has a common mode threshold comparator (with two thresholds for 200 and 400 Mbps), which is referenced to TpBias internally and is used for speed signaling. Speed shifting test is accomplished with the Vt drive level from the tester.

References

[1]   Mahoney, M., *Tutorial DSP-Based Testing of Analog and Mixed-Signal Circuits,* Los Alamitos, CA: IEEE Computer Society Press, 1987.

[2]   Souders, T. M., and D. R. Flach, "An NBS calibration service for A/D and D/A converters," *Proc. IEEE Int. Test Conf.*, 1981, pp. 290–303.

[3]   Max, S., "Fast, accurate and complete ADC testing," *Proc. IEEE Int. Test Conf.*, 1989, pp. 111–117.

[4]   Weimer, J., et al., "A rapid dither algorithm advances A/D converter testing," *Proc. IEEE Int. Test Conf.*, 1990, pp. 490–507.

[5]   Waterfall, G., and B. C. Baker, "Digital-servo and linear-regression methods test high-resolution ADCs," *EDN Magazine,* Fe. 1995, pp. 113–122.

[6]   Kimura, S., M. Kimura, and T. Nakatani, "A new approach for PLL characterization on mixed signal ATE," *Proc. IEEE Int. Test Conf.*, 1993, pp. 697–704.

[7]   Sunter, S., "Cost/benefit analysis of the P1149.4 mixed signal test bus," *Proc. IEEE Int. Test Conf.*, 1995, pp. 444–450.

[8]   Fasang, P. P., "Design for testability for mixed analog/digital ASICs," *Proc. Custom Integrated Circuits Conf.,* 1998, pp. 16.5.1–16.5.4.

[9]   Toner, M. F., and G. W. Roberts, "A BIST SNR, gain tracking and frequency response test of a sigma-delta ADC," *IEEE Trans. on Circuits and Systems II: Analog and Digital Signal Processing,* Vol. 42, Jan. 1995, pp. 1–15.

[10] Arabi, K., B. Kaminska, and M. Sawan, "On chip testing data converters using static parameters," *IEEE Trans. VLSI,* Sep. 1998, pp. 409–419.

[11] Teraoka, E., et al., "A built-in self test for ADC and DAC in a single-chip speech CODEC," *Proc. IEEE Int. Test Conf.*, 1993, pp. 791–796.

[12] Ohletz, M. J., "Hybrid built-in self test (HBIST) for mixed analog/digital integrated circuits," *Proc. European Test Conf.,* 1991, pp. 307–316.

[13] BISTMaxx white paper, "Solving the system-on-a-chip problem of design-for-test," Opmaxx web site, http://www.opmaxx.com/.

[14] LogicVision web site, http://www.lvision.com/.

[15] Rajsuman, R., "Testing a system-on-a-chip with embedded microprocessor," *Proc. IEEE Int. Test Conf.*, 1999, pp. 499–508.

[16] Cron, A., "IEEE P1149.4—Almost a Standard," *Proc. IEEE Int. Test Conf.*, 1997, pp. 174–182.

[17] Sunter, S., "Implementing the 1149.4 Standard Mixed-signal Test Bus," in *Analog and Mixed-Signal Test* (B. Vinnakota, Ed.), Englewood Cliffs, NJ: Prentice Hall, 1998.

[18] McDermid, J. E., "Limited access testing: IEEE 1149.4 instrumentation and methods," *Proc. IEEE Int. Test Conf.*, 1998, pp. 388–395.

[19] Lofstrom, K., "A demonstration IC for the P1149.4 mixed-signal test standard," *Proc. IEEE Int. Test Conf.*, 1996, pp. 92–98.

[20] Gasbarro, J. A., and M. Horowitz, "Techniques for characterizing DRAMs with a 500 MHz interface," *Proc. IEEE Int. Test Conf.*, 1994, pp. 516–525.

[21] Blancha, B., and L. Melatti, "Testing methodology for Fire Wire," *IEEE Design and Test of Computers*, vol. 16(3), July-Sep. 1999, pp. 102–111.

9

Iddq Testing

System-on-a-chip (SoC) ICs in deep submicron technologies present major challenges when attempting to implement Iddq testing. The problems are increased leakage current due to the increased number of gates and the increased subthreshold leakage of the individual transistors. While methods such as substrate bias and low temperature are used to reduce subthreshold leakage in the deep submicron technologies [1], partitioning methods are used to address the issue of increased leakage due to the enormous sizes of SoC designs. A detailed description on Iddq testing is given in [2, 3]; in this chapter only SoC-related issues are discussed.

9.1 Physical Defects

In any electrical circuit, opens and shorts are the fundamental physical defects. Some defects such as partial open and resistive bridging may not cause a gross failure but can cause a timing-related error or degraded reliability. Almost all studies of physical defects show that only a small fraction of defects can be modeled at the stuck-at level. The conclusions from these studies follow:

1. Wafer defects are found in clusters. These clusters are randomly distributed over the whole wafer. Every part of the wafer has an equal probability of having a defect cluster.

2. Any part of a diffusion, polysilicon, or metal line may have an open fault. Any contact between any two layers may be open.

3. Bridging may occur between any two electrical nodes, whether they belong to one layer or different layers. Bridging among multiple nodes is equally likely.

4. Only a small percentage of bridging and open faults can be modeled at the stuck-at level. The actual distribution varies and depends largely on the technology and fabrication process.

To understand Iddq behavior, bridging and open defects are discussed in separate subsections.

9.1.1 Bridging (Shorts)

Some examples of bridging defects are shown in Figure 9.1 [2, 4–6]. In Figure 9.1(a), seven metal lines are bridged together due to an unexposed photoresist; in Figure 9.1(b), four metal lines are bridged together due to the presence of a foreign particle; in Figure 9.1(c), a few lines exhibit bridging and opens due to a scratch on the mask; in Figure 9.1(d), a 1-μm-size killer defect causes a catastrophic short; in Figure 9.1(e), a metallization defect causes a single bridging between two Al lines; and in Figure 9.1(f) an inter-layer short is shown. The cause of defects in each of these examples is different, but the results are either bridging or opens.

The simulation and modeling of such defects can be done by *inductive fault analysis,* examples of which are given in Figure 9.2 [4]. Circuit schematics are also included in Figure 9.2 to illustrate the effect of these defects. Figure 9.2(a) shows a spot defect causing extra polysilicon. The corresponding transistor-level and gate-level schematics show that this defect can be modeled at the stuck-at level. However, the examples of spot defects in Figures 9.2(b)–(d) are not modeled at the stuck-at level. In Figure 9.2(b), an extra transistor is formed; in Figure 9.2(c), two inverters convert into a NAND gate; and in Figure 9.2(d), circuit topology is changed, which results in a different Boolean output.

Many papers are available on bridging faults that attempt to detect bridging by logic testing in a voltage environment [7–9]. In [8] and [9], analytical models were developed to explain the bridging behavior.

The behavior of bridging can be explained by a potential divider rule as shown in Figure 9.3(a). The outputs of two logic elements are indicated by

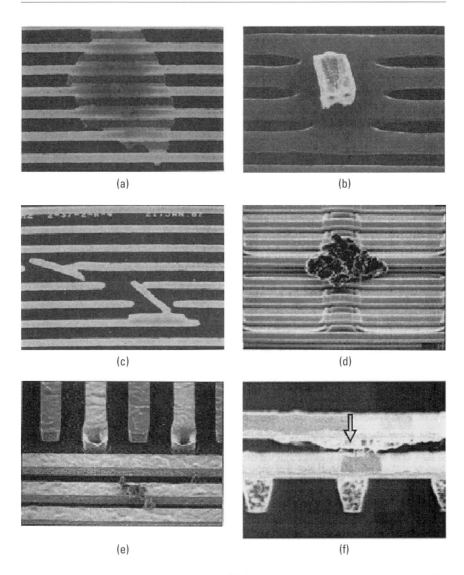

(a)

(b)

(c)

(d)

(e)

(f)

Figure 9.1 Examples of bridging defects: (a) Shorting of seven metal lines caused by unexposed photoresist [4]. (b) Shorting of four metal lines by a solid-state particle on the metal mask [4]. (c) Shorts and breaks of metal lines caused by a scratch in the photoresist [4]. (d) Defect short among multiple metal lines by a metallization defect of 1-μm in size [5]. (e) Short between two Al lines due to metallization [5]. (f) Interlayer short between two Al interconnects in 0.5-μm technology [6]. (© IEEE 1987, 1994, 1997, 2000. Reproduced with permission.)

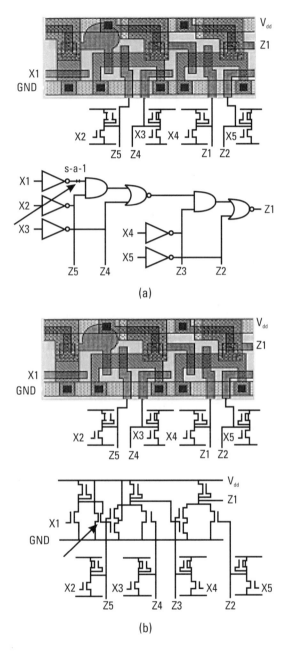

Figure 9.2 Examples of spot defects and their effect on the circuit: (a) Spot defect causing extra polysilicon results in an s-a-1 fault. (b) Spot defect causing extra
active region results in an extra transistor and a short to the Vdd line. (From
[4], © IEEE 1987. Reproduced with permission.)

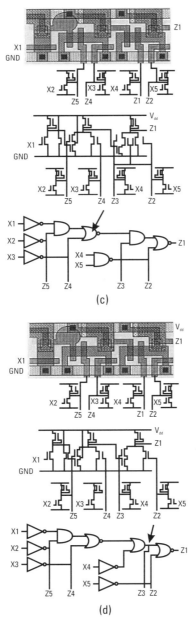

Figure 9.2 (Continued) Examples of spot defects and their effect on the circuit: (c) Spot defect causing extra polysilicon results in the bridging of inverter outputs consequently transforming them into a NAND gate. (d) Spot defect causing extra polysilicon results in transistor bridging that changes circuit topology and Boolean output. (From [4], © IEEE 1987. Reproduced with permission.)

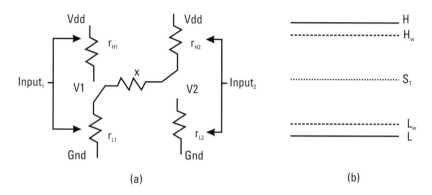

Figure 9.3 (a) Simple potential divider model for bridging. (b) Definitions of voltage levels. (From [8], © IEEE 1986. Reproduced with permission.)

subscripts 1 and 2, and the bridge resistance is shown as x. Let the resistance that connect the L (H) node to ground (Vdd) be r_L (r_H). If the worst-case (min) output H voltage is H_w, the H-level noise margin is N_H, the worst case (max) output L voltage is L_w, the L-level noise margin is N_L, and the switching threshold voltage is S_T (the definitions are shown in Figure 9.3b). The voltage at the output nodes V_1 and V_2 can be given as $V_1 = \text{Vdd}[r_{L1}/(r_{L1}+x+r_{H2})]$ and $V_2 = \text{Vdd}[(r_{L1}+x)/(r_{L1}+x+r_{H2})]$.

Bridging is not consequential if $x \rightarrow \infty$, giving $V_1 = 0$ and $V_2 = \text{Vdd}$. For a low-resistance bridging (hard short), $x \rightarrow 0$, which gives $V_1 = V_2 = \frac{1}{2}\text{Vdd}$, when r_L and r_H are equivalent.

This potential divider model indicates that in the case of high-resistance bridging the voltage at the defect site may or may not cause a logical error depending on the bridge resistance. A hard short or low-resistance bridging will cause $\frac{1}{2}\text{Vdd}$ at the defect site. The $\frac{1}{2}\text{Vdd}$ implies indeterminate logic value at the defect site. Thus, even if it is sensitized by logic testing, fault detection is not likely because indeterminate logic value at the defect site will become either logic 1 or 0 during fault propagation.

Bridging resistance is important in defect detection. The majority of bridging defects show low resistance; only about 20% of the bridges are of significant resistance [10, 11]. High-resistance bridging does not result in failure at the time of testing, and affects only the noise margin (the output between H_W and L_W may not necessarily cause a logical error). However, these defects significantly degrade device reliability.

Fortunately, a large number of bridging defects (including a large number of resistive bridging) can be detected by Iddq testing. As is clear from

Figure 9.3, in the presence of bridging, a conduction path is formed from Vdd to Gnd. Subsequently, the circuit dissipates large current through this path and, thus, simple monitoring of the supply current can detect bridging. This detection does not require any fault propagation; the fault propagation is automatic through the power supply.

9.1.2 Gate-Oxide Defects

The various defects and reliability issues of gate oxides are well known. These defects include pinholes and micropores, dendrites, trapped charges due to hot carriers, nonstoichiometric $Si–SiO_2$ interface, and direct short to diffusion. Examples of a gate-oxide short to N^+ diffusion and a gate-oxide pinhole are shown in Figure 9.4 [12, 13].

Some of these defects occur during oxidation or other thermal processes, while other defects may occur due to electrostatic discharge or overstress (ESD/EOS).

In today's $0.25\text{-}\mu\text{m}$ technology (and below), a gate oxide of 50 to 60Å thickness is used for logic MOSFETs and as low as 35 to 40Å for devices such as EEPROMs and flash ROMs. Although, gate-oxide thickness is tightly controlled (most fabs target ±2 to 3Å), the smallest variation in thickness increases the possibility of a defect occurring. For example, the lower thickness region may cause Fowler-Nordhiem tunneling and, in extreme cases, avalanche breakdown during voltage stress tests. ESD/EOS-induced breakdowns are also very common in such thin oxides.

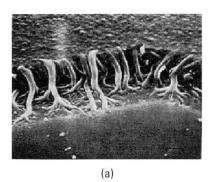
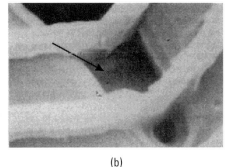

(a)   (b)

Figure 9.4 Examples of gate-oxide defects: (a) gate-oxide short to N+ diffusion [12]. (b) Gate-oxide pinhole causing cell-to-word line leakage in a memory [13]. (© IEEE 1986, 1994. Reproduced with permission.)

Gate-oxide reliability issues and breakdown mechanisms have been extensively reported in the literature. In the majority of cases, gate-oxide defects cause reliability degradation such as change in transistor threshold voltage (Vt) and increased switching delay; only in some cases (such as avalanche breakdown and subsequent short) it causes a logical failure. Some papers also present very elaborate models for gate-oxide shorts and defects [14–16]. However, in general, logic testing does not detect gate-oxide defects [17, 18], primarily due to difficulties with fault-effect propagation. Iddq testing, on the other hand, is very effective in detecting these defects because they cause high current dissipation in the circuit.

9.1.3   Open (Breaks)

Analysis of physical defects from fabs and from inductive fault analysis suggests that approximately 40% of defects are open [19]. The open defects are much more difficult to detect by logic testing and, in fact, Iddq testing does not necessarily detect them. Examples of open defects include line open and line thinning (it may or may not be a partial open at the time of testing) [4]. Figure 9.5(c) shows yet another example of a defect that caused both an open and a short [20].

In CMOS circuits, many opens will cause sequential behavior. The difficulty of detecting an open fault can be illustrated by a simple example. Figure 9.6 shows a two-input NOR gate with an open in drain-to-output connection of nMOS with input B. All four test vectors are shown in the top part of the adjacent table. As marked in the table, the vector AB = 01 is the only vector that sensitizes this open. However, during this vector, in the presence of the open, output of the gate is in high impedance. Hence, the vector before the sensitization vector defines the logic value at the output. If it is AB = 10 or 11, the output remains at 0 and open is not detected. To detect this open, the necessary sequence of patterns is AB = 00, 01.

A large number of papers have been written about detecting opens by logic testing using two or multiple-pattern tests [21–24]. A large number of reports also show that due to differences in delays along various paths and charge sharing among internal nodes of a gate, two or multiple-pattern tests may become invalidated. These papers also suggest that testable designs be used [23–25], while another set of papers suggests that robust two or multiple-pattern test sequences be used [26, 27]. Besides causing logical failures, open defects can also cause timing-related errors; in particular, open-gate defects are very sensitive to capacitive coupling. Thus, suggestions have

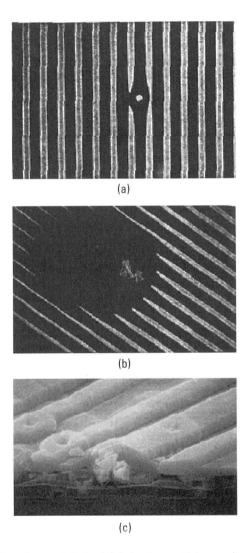

(a)

(b)

(c)

Figure 9.5 Examples of open defects: (a) A foreign particle causing a line open and a line thinning [4]. (B) A contaminating particle causing seven line opens [4]. (c) SEM picture of a defect that caused an open in metal 2 and a short in metal 1 [20]. (© IEEE 1987, 1993. Reproduced with permission.)

been made to detect open defects using two or multiple-pattern tests developed to detect delay faults [28, 29].

Both testable designs and two or multiple-pattern sequences (including robust sequences) do not provide a practical solution to detect open defects.

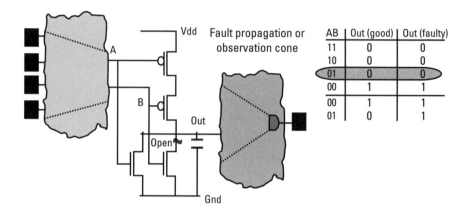

AB	Out (good)	Out (faulty)
11	0	0
10	0	0
01	0	0
00	1	1
00	1	1
01	0	1

Figure 9.6 Example illustrating the difficulty in detection of an open fault. During AB = 01, the node marked "Out" is in a high-impedance state, but from the chip's pin, it still appears at "0." The two-pattern sequence AB = 00, 01 detects this fault.

Testable designs require redesign of the standard cell library and massive routing of additional global signals, and robust test sequences require elimination of all possible glitches in the circuit, so these are difficult to identify. Although a few companies have developed internal tools to generate two-pattern tests for opens, neither universities nor the EDA industry have so far been successful in developing a suitable ATPG tool and pattern sequencing mechanism to deterministically detect open defects in a cost-effective manner by logic testing.

Unfortunately, Iddq testing is also not very effective for open defects. Although some reports suggest that Iddq testing can be used to detect opens [29, 30], such detection is highly subjective to the cell design style and topology of the circuit. This behavior is clearly explained by a detailed electrical model of opens, such as that given in [31]. In [32], the effectiveness of Iddq testing was evaluated for open defects by intentionally fabricating an open, and it was concluded that Iddq testing is not very effective. In a simple example, such as that given in Figure 9.6, there is no current dissipation when the two-pattern test (such as AB = 01, 10) is used, which also fails to detect open defects in the logic testing. Even when the two-pattern test (such as AB = 11, 10) that detects open defects in logic testing is used, there is no static current dissipation in the circuit. Hence, in both situations, Iddq testing remains ineffective to detect this open.

9.1.4   Effectiveness of Iddq Testing

With the background of Sections 9.1.1 to 9.1.3, it is clear that Iddq testing is very effective for defects such as bridging and gate-oxide shorts, but not as effective for opens. Because it does not verify the functionality, it is used as a supplemental test to functional testing. The same is also true for reliability testing such as stress testing and burn-in. In both cases, Iddq testing adds significant value by improving the quality of the test at a very low cost. Since its infancy, studies have shown that a large percentage of defects can be detected by Iddq testing; however, in almost all studies, there are also failures identified by either functional or scan-based testing that remain undetected during Iddq testing [33–35]. This relationship is qualitatively shown in Figure 9.7 [36].

In Figure 9.7, A2 faults are undetected faults in logic testing; A3 are the faults detected by logic testing alone; A4 are undetected faults in Iddq testing; A5 are the faults detected by both logic and Iddq testing; A6 are faults undetected in logic testing but detected in Iddq testing; A7 are faults detected by Iddq testing alone; A8 are Iddq-testable faults that remain undetected in Iddq testing; and A9 are faults undetected in both logic and Iddq testing.

The net benefit of Iddq testing is (A6 + A7). The A6 shows logic test generation effort is minimized and A7 is additional defect coverage. Even for

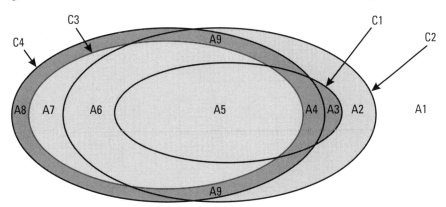

Figure 9.7   Relationship of the effectiveness of Iddq and logic testing. The A1's are faults not detected by either Iddq testing or logic testing; C1 faults are detected by logic testing; C2 faults are all logical faults; C3 faults are detected by Iddq testing; C4 faults are all Iddq-testable faults. (From [36], © IEEE 1991. Reproduced with permission.)

the Iddq test set of just 10 to 100 vectors, (C4 − C3) is generally small. For a design when (C2 − C1) is large (for example, functional coverage is only 60%), the overall coverage (C1 ∪ C3) can still be made sufficiently high to be acceptable by logic plus Iddq testing without incurring significant cost. There will always be some defects that cannot be detected by Iddq testing, as shown by A2 + A3 in Figure 9.7.

This conclusion was also drawn by experimental studies conducted to evaluate the effectiveness of various test techniques [37–41]. In all studies, Iddq testing detected a majority of defects—however, a few defects undetected in Iddq testing were detected in logic testing. At the same time, some defects undetected in logic testing were detected in Iddq testing. A summary of data from a Sematech study is given in Figure 9.8 [40].

The data in Figure 9.8 are typical of studies of this nature, which compare the effectiveness of various testing methods. As marked in Figure 9.8, 36 devices passed the Iddq test, but failed every other test. On the other hand, as marked on the first row, 1463 devices failed the Iddq test (Iddq threshold limit 5 μA), but passed every other test. This particular study was based on a 116K-gate standard cell graphics controller chip designed in IBM Phoenix CMOS4LP technology. This technology has 0.45 μm L_{eff} and 0.8 μm L_{drawn}. A total of 20,000 dies were tested at wafer sort, out of which

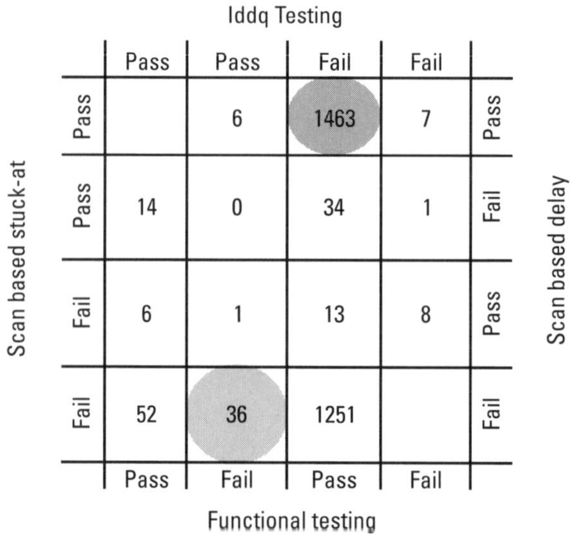

Figure 9.8 Summary of data from Sematech study. (From [40], © IEEE 1997. Reproduced with permission.)

4000 were more carefully tested at the package level. The chip characteristics were [42]:

- 249 signals I/Os, flip-chip/C4 wafer contacts, 304-pin C4FP package;
- Full-scan design, 5280 LSSD latches, boundary scan;
- Functional speed 40/50 MHz, function being bus interface controller;
- Designed for Iddq testing, typical Iddq < 1 μA;
- Scan based stuck-at test set with 99.7% stuck-at fault coverage; functional test set with design verification vectors of 52% coverage; scan-based delay testing with >90% transition fault coverage; Iddq test set with 195 vectors.

All packaged devices also experienced a minimum of 6 hours of burn-in, while a large sample had 150 hours of burn-in. Identical tests were used at wafer sort, pre- and post-burn-in package level tests.

Detailed failure analysis was done to identify the defects on a number of devices that failed one or more tests. Iddq testing was found to detect various kind of bridging (not detected by any other test), including low- and high-resistance bridging (75Ω, 194Ω, 1.63 kΩ, 184 kΩ, 194 kΩ, 340 kΩ). However, an open resulting in a floating gate was not detected by Iddq testing; it was, however, detected by scan-based stuck-at testing.

9.2 Iddq Testing Difficulties in SoC

While Iddq testing is a very desirable test method, its implementation on SoC designs presents considerable challenges. The major difficulty in performing Iddq testing on embedded cores for SoC designs is the enormous size of the design. Even the simple SoC designs of today contain more than a million transistors, and designs with 10 to 15 million transistors are becoming quite common. In these large designs, the summation of leakage current (I_{off}) of all transistors becomes too large to distinguish between faulty and fault-free chips. Most of the SoC designs contain multiple power supplies (3.3V, 2.5V, 1.8V, and so on) and Iddq testing is done on one power supply at a time. However, in most cases, the majority of the digital portion consisting of multiple cores is powered by one power supply and subsequently shows too large a leakage current for Iddq testing to be feasible.

Another way to look at this problem is to recognize that the size of embedded cores today is in the range of 20 to 200K gates. When four or five cores on chip are powered by one power supply, the power supply loading becomes 100K to a million gates. The combined leakage of all four or five cores makes it impossible to perform Iddq testing.

The theory of Iddq testing is based on estimating the fault-free current in the circuit and then setting a limit (popularly called an *Iddq threshold*), above which a circuit is considered faulty. While some research papers have been published on the methodology and tools used to estimate fault-free current, in industry it is based upon the measurements of large numbers of devices. Although the number varies, in general, for digital circuits of 0.5-, 0.35-, and 0.25-μm technologies, a number close to 1 μA is considered to be fault free and any number from as low as 10 μA to as high as 100 μA is considered to be threshold. Due to the law of large numbers, the distribution of measured current over a large number of ICs is expected to be Gaussian. Due to statistical variations, ICs up to (mean + 3σ) are considered fault-free. A limit higher than (mean + 3σ) is assumed for the Iddq threshold, above which ICs are considered faulty. This concept is illustrated in Figure 9.9 (for illustration, the distribution of faulty current is also assumed to be Gaussian).

The separation between the distribution of faulty and fault-free current is:

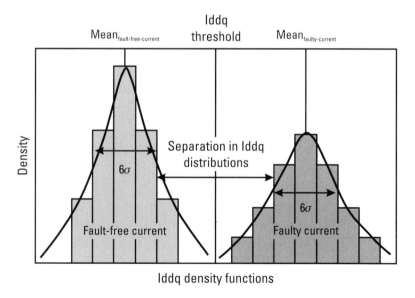

Figure 9.9 Representation of fault-free and faulty Idd density functions.

$$\text{Iddq}_{\text{fault}} - \text{Iddq}_{\text{fault-free}} = (\text{average Iddq}_{\text{fault}} - 3\sigma_{\text{Iddqfault}}) -$$
$$(\text{average Iddq}_{\text{fault-free}} + 3\sigma_{\text{Iddqfault-free}}) \qquad (9.1)$$

When the density functions of fault-free and faulty currents are separated from each other, the clear distinction between the good and the faulty IC can be made easily. However, as the technology shrinks, the subthreshold leakage (I_{off}) per transistor increases, and the number of gates in the SoC design increases, the total leakage in the circuit increases. Thus, the distribution of fault-free current flattens and moves toward the distribution of faulty current. When these two distributions overlap, it becomes impossible to clearly distinguish between the good and the bad IC.

The I_{off} represents steady-state leakage of a transistor; Iddq in an IC can thus be approximated as the summation of all leakage:

$$\text{Iddq} = (\text{normalized number of nMOS.} \ [n\text{-}I_{\text{off}/\mu m}]) +$$
$$(\text{normalized number of pMOS.} \ [p\text{-}I_{\text{off}/\mu m}]) \qquad (9.2)$$

Equation (9.2) identifies two parameters that affect Iddq in the circuits: (1) I_{off} and (2) number of transistors. The normalized number of nMOS and pMOS is used above because transistors of various sizes are used within a design.

Let's consider the first parameter, subthreshold leakage or I_{off}. The problem becomes very clear from the numerical data of pMOS and nMOS I_{off} from 0.5-, 0.35-, 0.25-, and 0.18-μm technologies. The characteristic data for various technologies (with I_{off} values) are given in Table 9.1.

Table 9.1
Characteristics of Various Technologies and Range of I_{off} Data

Tech (μm)	Vdd (V)	T_{ox} (Å)	V_t (V)	I_{off} (pA/μm)	
				NMOS	PMOS
0.8	5	150–100	0.8–0.7	0.01–0.05	0.005–0.02
0.6	5–3.3	100–80	0.75–0.65	0.05–0.5	0.01–0.2
0.5	5–2.5	90–70	0.7–0.6	0.1–2	0.1–1
0.35	3.3–2.5	80–60	0.65–0.55	0.5–10	0.1–10
0.25	3.3–1.8	70–50	0.6–0.5	6–60	0.5–24
0.18	2.5–1.8	55–35	0.55–0.45	40–600	20–300

Table 9.1 provides the range of parameters of various manufacturers for a specified minimum feature size. The data in Table 9.1 clearly shows that I_{off} has increased 4 to 5 orders of magnitude from 0.8- to 0.18-μm technology.

The reason for this increase in I_{off} can be understood by transistor's V_G versus I_D transfer curve, as shown in Figure 9.10 [1]. The I_{off} is measured at $V_G = 0V$, and for this transistor, it is 20 pA/μm for the saturated region and 4 pA/μm in the linear region. The subthreshold slope S_t (V_G versus I_D in the weak inversion region) is about 80 mv/decade of I_D. S_t is a function of gate-oxide thickness and the surface doping adjusted implant. The change in S_t is minimized by T_{ox} scaling and improved doping profiles. A $S_t \geq 100$ mV/decade indicates a leaky device, while a lower value results in low I_{off} for a given threshold voltage. S_t has been reported to change from about 75 in 0.35-μm technology to about 85 in 0.25-μm technology.

Another reason for increased I_{off} is drain-induced barrier lowering (DIBL) and gate-induced drain leakage (GIDL). DIBL moves the curve up and to the left as V_D increases while GIDL current shows up as a hook in the transistor I_D versus V_G curve (Figure 9.10). In general, the nature of I_{off} with technology shrink is given as shown in Figure 9.11.

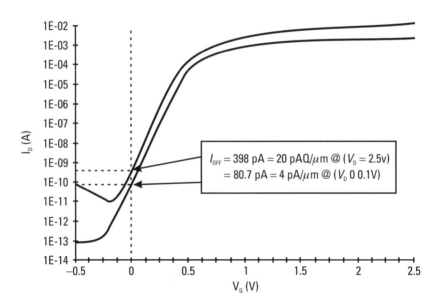

Figure 9.10 Log I_D versus V_G at saturated bias ($V_D = 2.5V$) and linear bias ($V_D = 0.1V$), for a 20 × 4-μm nMOS transistor. (From [1], © IEEE 1997. Reproduced with permission.)

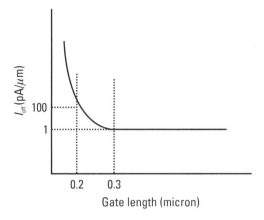

Figure 9.11 Nature of I_{off} with geometry shrink.

As shown in Figure 9.11, I_{off} has been less than 1 pA/μm for a gate length 0.35 μm or larger and it starts to increase exponentially at about 0.25 μm. The discussion above provides the underlying reasons for increased I_{off} and concern for Iddq testing in deep submicron technologies. It is also clear from the above discussion that if the I_D versus V_G curve in the linear and subthreshold regions is moved down to the right, the problem of high leakage will go away and, hence, Iddq testing will continue to be useful.

Two mechanisms were proposed in [1]: (1) reduced temperature and (2) substrate bias. A mathematical model to explain the effect of back bias on subthreshold current is given in [43]. The quantitative data to illustrate the effect of temperature and substrate bias is shown in Figures 9.12(a) and (b), respectively.

The dramatic reduction in I_{off} is clear from Figure 9.12. As an example, 42 pA I_{off} at room temperature can be reduced to about 9 pA at 0°C—a reduction factor of about 4.5. Similarly, 9.6 nA I_{off} at V_{sub} = 0V can be reduced to about 2 pA at V_{sub} = −4V—a reduction factor of about 4400. It should be noted that beyond a certain point, a further decrease in V_{sub} increases I_{off}. Lowering both V_{sub} and V_D, however, reduces I_{off}. Using these methods, a reduction factor as high as 60,000 has been reported.

The second parameter in (9.2) is the number of gates in the circuit. The increasing number of gates in an IC also results in an increase in the leakage current. With both increasing circuit size (number of gates) and increased I_{off}, the distribution of the fault-free current flattens and approaches the Iddq threshold. Changing the threshold limit to a higher number does not solve the issue because change in fault-free and faulty

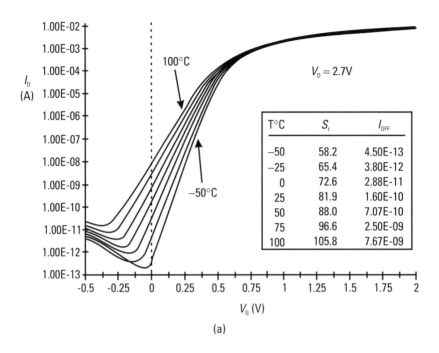

T°C	S_t	I_{OFF}
−50	58.2	4.50E-13
−25	65.4	3.80E-12
0	72.6	2.88E-11
25	81.9	1.60E-10
50	88.0	7.07E-10
75	96.6	2.50E-09
100	105.8	7.67E-09

(a)

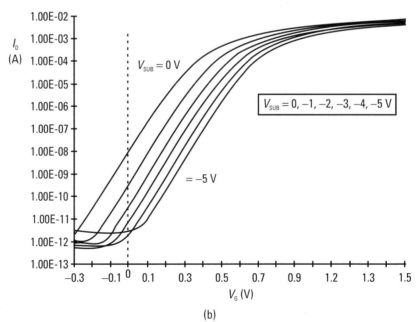

(b)

Figure 9.12 The effect of (a) temperature and (b) substrate bias on linear and subthreshold regions. (From [1], © IEEE 1997. Reproduced with permission.)

current due to a defect becomes minuscule and unidentifiable with higher leakage in the circuit. In this case, the distributions of fault-free and faulty current appear to be merged. Thus, the distinction between fault-free and faulty IC becomes blurred and Iddq testing becomes less and less useful in identifying a defect. To overcome the problem of a large number of gates, proposals have been made to partition the circuit and to perform Iddq testing on one partition at a time. We describe these methods in the next section.

9.3 Design-for-Iddq-Testing

In [44–47], a design methodology is presented that provides a global control signal to switch off static current dissipating logic. This design uses a special buffer and a dedicated pin to control this global signal during Iddq testing as well as during normal operation. The basic design is shown in Figure 9.13 [44]. During normal mode in Figure 9.13—due to the pull-down transistor of special buffer (Iddtn buffer)—the control signal (Iddtn-ring) is always "1,"

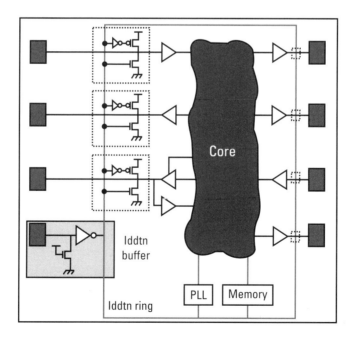

Figure 9.13 ASIC design that allows the static current dissipating logic to be switched off during Iddq testing. (From [44], © IEEE 1995. Reproduced with permission.)

keeping the pull-down transistors ON at the signal pins. For Iddq testing, a "1" is applied to this dedicated pin, which brings Iddtn-ring to "0" and switches off all pull-down transistors on the signal pins. The connections to the PLL, memories, and other switchable internal logic occur via an inverter (not shown in Figure 9.13) connected to the gate of a large pMOS transistor in the power supply path.

The hardware overhead appears to be negligible in this design. However, the major limitation is that it requires one dedicated pin to control the global signal (Iddtn-ring).

One can extend this method to SoC designs by one power supply control signal per core; thus, overcoming the problem of large leakage of multiple cores during Iddq testing. However, if this concept is used as is in SoC design, then one pin per core is needed to control the power supply to the core. This means that for SoC designs with four or five cores, four to five pins will be required for Iddq testing. In the majority of cases, even one pin at the chip level is very costly, so four or five pins just for Iddq testing is extremely expensive.

To overcome the requirement of a dedicated pin, JTAG boundary-scan-based control is used [46]. For this purpose, a flip-flop is used to control the logic of the global Power_Down control signal. During normal mode, the flip-flop is set to "1"; in the Iddq test mode, it is set to "0." A new private instruction, "Power_Down," is implemented along with other mandatory and private JTAG boundary-scan instructions (EXTEST, INTEST). In order to perform Iddq testing, the Power_Down instruction is loaded into the boundary-scan instruction register. This instruction is decoded and it provides a logic "0" to the flip-flop controlling the Power_Down control signal. With the Power_Down control signal at 0, the TAP controller is kept in the RunTest/Idle state for the duration of Iddq testing. Hence all the circuits connected to the Power_Down control signal switch off and remain off. After Iddq testing is completed, the TAP controller is brought back to the reset state. This resets the flip-flop to "1" and thus brings the Power_Down control signal back to 1; the circuit subsequently returns to normal operating mode. The detailed design is shown in Figure 9.14 [46].

As mentioned earlier, some type of partitioning method is required for Iddq testing in embedded core-based system chips. The large size of the design demands it. Rather than physically partitioning the circuit, in this design Power_Down control signals are used to selectively switch off portions of the system chip. Specifically, multiple cores and other static current dissipating logic are switched off and Iddq testing is performed on one core at a time.

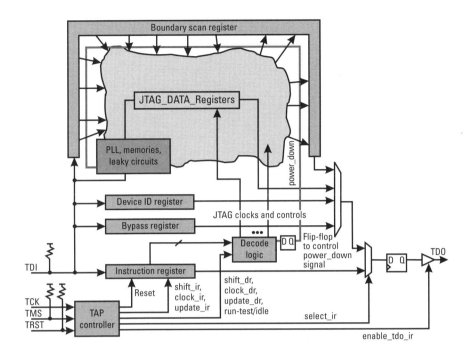

Figure 9.14 Implementation of a global Power_Down control signal through a TAP controller. (From [46], © IEEE 1999. Reproduced with permission.)

The method shown in Figure 9.14 is extended to control the power supply to the individual cores during Iddq testing. The concept is illustrated in Figure 9.15, which contains three cores and one large embedded memory. All the cores in this figure contain a wrapper around them, which is used for accessing the core during logic testing.

In this design, four private instructions are used, Power_Down_A, Power_Down_B, Power_Down_C, and Power_Down_Main. The nomenclature is that these instructions allow Iddq testing on the block/core identified in the instruction. For example, the Power_Down_A instruction cuts off the power supply of core B, core C, glue logic, PLL, memory, and any static current dissipating logic in core A. Thus, it allows Iddq testing on core A. A 4-bit register is used which provides the necessary logic values for Power_Down control signals to selectively perform Iddq testing on any one core or glue logic. When any of the Power_Down_A, Power_Down_B, or Power_Down_C instructions are loaded onto the boundary-scan instruction register, one Power_Down signal is kept at 1 while all other Power_Down

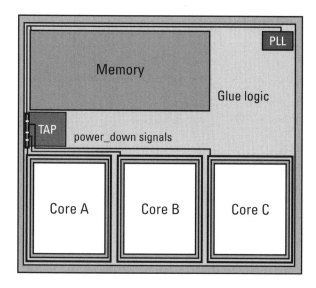

Figure 9.15 Implementation of Power_Down control signals to perform Iddq testing on embedded cores in SoC.

signals are set to 0. The Power-Down control signals at 0 cuts off the power supply of the respective blocks.

The decoding of the Power_Down_Main instruction is slightly different. It sets all Power_Down signals to 0, and therefore switches off all individual cores, PLL, memory, and other static current dissipating logic in glue logic. Thus, it allows Iddq testing on glue logic.

Note that the hardware overhead in this design is negligible. Because Power_Down control signals are wires, they do increase routing but do not increase die size; and no dedicated pin is required to control Power_Down signals. It is also important to note that when individual components (all cores, PLL, memory sense-amps, pull-up/pull-down, dynamic logic, and so on) are designed with a power-down control signal, a consistent naming convention significantly helps in the design. With a consistent naming convention, no additional design effort is needed for the implementation of global control signals. During placement and routing, the router sees the same names throughout the chip—from the TAP to the cores—and connects them appropriately to automatically form the Power_Down control signals.

The above method (demonstrated in Figures 9.14 and 9.15) is also applicable to the IEEE P1500 standard (under development) for embedded core testing. Instead of an extra register in the TAP controller for controlling the Power_Down signals, an equivalent register can be implemented in the control scan path (CSP) of the P1500 with exactly the same functionality as given in Figure 9.14 and Figure 9.15 [48].

Besides the above design feature, the choice of package is also considerably useful in Iddq testing. A segmented Vdd plane allows a natural partitioning of the power supply at chip level. Such a package with a four-segment Vdd plane is shown in Figure 9.16. Although each segment has the same voltage, logically we can consider there to be four power supplies: Vdd1 = Vdd2 = Vdd3 = Vdd4.

During testing, each segment is connected to a different tester power supply or independent source. Iddq testing is performed on one segment at a time. Although all four segments are powered on for chip operation, the leakage in the circuitry connected to each segment is isolated from each other and limited to an independent source. Thus, the leakage as observed at any one segment is significantly less than the cumulative leakage of the whole chip, if the whole SoC is powered by one source. This partitioning method allows for very good resolution in Iddq testing and it also avoids interference due to background leakage without any DFT effort, area overhead, or any other type of penalty.

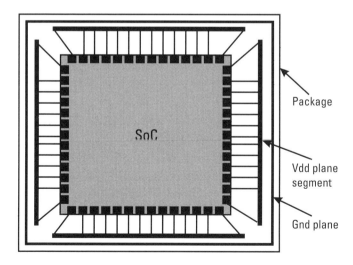

Figure 9.16 SoC package with segmented Vdd plane that allows natural partitioning and Iddq testing on one segment at a time. Signal and Gnd connections are not shown in the figure.

9.4 Design Rules for Iddq Testing

In addition to the above design, a number of other design rules are necessary to avoid unwanted high Idd states in the SoC IC. These rules are summarized as follows [46]:

1. All circuits need to be properly initialized. Full circuit initialization is a fundamental requirement for Iddq testing. Besides using set/reset of flip-flops or a dedicated signal, full scans or partial scans can be used very effectively to initialize the circuit. It means that all flip-flops (registers) must be placed in a known state, such as flip-flops within a core when the core is tested or glue logic when glue logic is tested. This initialization can be done via a set/reset signal or through scan operation.

2. All static current dissipating logic within a core or glue logic must be switched off using power-down control signals. This includes all memory sense amps, any dynamic logic, asynchronous logic, pull-up/pull-down resistors, special I/O buffers, and any analog circuitry.

3. All circuits including individual cores and glue logic must be stable at the strobe point. There should be no pending events during Iddq testing either within a core, from one core to another core, or at the chip level.

4. All inputs and bidirectional pins of all individual cores as well as pins at the chip level must be at a deterministic 0 or 1.

5. At the core level, if an input, output, or bidirectional pin is pulled up, it should be at a logic 1, connected to Vdd through a pMOS; if it is pulled down then it should be at a logic 0, connected to Gnd through an nMOS. All pull-up and pull-down transistors at the chip I/Os must be switched off.

6. All primitive nets within a core or glue logic with single driver must be checked for the following: (a) all nets are at either logic 0 or logic 1; (b) if a net is at x, either the driver should not be tri-stateable or it should not be driven by a tri-stateable gate whose enable pin is active; (c) any net should not be at a high-z state. These conditions are necessary to ensure that there is no internal bus conflict or floating nodes during Iddq testing.

7. When primitive nets are driven by multiple drivers, the following must be checked: (a) net should not be driven both 1 and 0 at any given time; (b) net should not be driven to $0/x$, x/x, $0/0/x$, or $1/x$.

In all of these conditions there is a potential conflict on the net; (c) net should not be driven to x/z, $z/z/x$, or $z/x/x$ —in these situations, net is either potentially floating or has a conflict.

8. All nets within a core and glue logic should be checked to ensure that there is no weak value feeding to a gate during Iddq measurement. Similarly, node feeding to a gate during Iddq measurement should not have a degraded logic value.

9. Special circuit structures should be avoided as much as possible. When such structures are unavoidable, a Power_Down control signal should be used to switch off these structures during Iddq testing. The examples of such structures are gate and drain/source of a transistor be driven by the same transistor group; feedback and control loops within one transistor group; substrate connection of nMOS being not at Gnd, and a substrate connection of pMOS being not at Vdd.

10. A standard cell library that contains components with low-power switches and use of a separate power supply for digital logic, I/O pad ring, and analog circuit is also helpful. In this situation, Iddq testing on digital logic can be done easily.

9.5 Iddq Test Vector Generation

A number of tools have been developed internally by corporate CAD and EDA companies as well as by universities. While most of the tools select vectors from the functional test set based on the user's defined constraints, some tools also contain an Iddq ATPG. Following are the general characteristics of these tools:

1. Ability to obtain Iddq vectors for the stuck-at coverage or toggle coverage. The input for this option is usually the design RTL netlist and testbench or gate-level netlist and testbench. The vectors are selected from the testbench (Verilog or VHDL) vectors.

2. Ability to obtain Iddq vectors for pseudo stuck-at (PSA) coverage. (PSA fault model is similar to stuck-at fault model. However, Iddq-oriented fault simulation does not require fault-effect propagation through the whole circuit; it is propagated only through one gate and observed through the power supply. Thus, the coverage numbers in such simulation are considered to be pseudo stuck-at [33]). The input for this option is gate-level netlist and testbench.

3. Ability to obtain Iddq vectors for toggle coverage. The input for this option is again the RTL netlist or gate-level netlist.

4. Ability to generate Iddq vectors for the bridging fault model. The input for this option is either RTL netlist and testbench or the gate-level netlist and testbench. This option is generally associated with an Iddq ATPG, which generates tests by providing opposite logic values on two lines (line 1 and line 0).

5. Ability to generate Iddq vectors for physical defects. The input for this option is layout (GDSII) and min/max particle size (defined by the user). This option is associated with an Iddq ATPG. The tools that use this option are experimental; no commercial tool is yet available with this option.

In addition, various simulation and timing requirements must be met by a tool if it is to correctly select/generate Iddq vectors and a fault coverage report. Generally, an Iddq test tool (ATPG or vector selection tool) is linked with a Verilog/VHDL simulator. This link is established before the design simulation run.

The user-defined constraints for Iddq vectors are passed through a simulation control file that is similar to a configuration file in Verilog simulation. The constraints and link to design simulation let the tool identify any Iddq rule violation. A user can waive any of the violations after reviewing them. However, because of these violations, many vectors are dropped from being a candidate in the Iddq test set (ideally, if there is no violation, all functional vectors can become Iddq vectors). The remaining vectors are "qualified Iddq vectors"; these vectors are fault graded under a fault model specified by the user.

A table is created that lists the vectors as well as detected faults; the format of this table varies for various tools. Based on user-specified constraints, the necessary number of vectors is taken from this table to form the Iddq test set. For example, the Iddq test set may be the 25 best vectors or n vectors that provide 90% coverage. This Iddq test set with a fault coverage number is reported in a separate file.

The vector selection process is generally based on one of the following two procedures:

Procedure Vector Selection (A):

Step 1:   Select a test vector that detects the maximum number of outstanding faults under the user-selected fault model. Add this vector into the list of selected vectors.

Step 2: Remove all faults detected by the selected vector from consideration.

Step 3: If number of selected test vectors exceeds the user's defined limit, exit.

Step 4: Repeat steps 1 to 3.

Step 5: Provide fault coverage by the selected vectors.

Procedure Vector Selection (B):

Step 1: Count the number of test vectors that detect each fault.

Step 2: Select all faults that are detected by the minimum number of test vectors. Mark the test vectors that detect these faults.

Step 3: Choose a test vector for the set of test vectors in step 2, which detects the maximum number of uncaught faults.

Step 4: Repeat steps 1 to 3 until all faults are detected.

The vector generation process targeted for bridging faults is generally as follows:

Procedure Vector Generation for Bridging:

Step 1: Let nodes j and k be the electrical nodes on two ends of a fault. Find the set N_j^0 of input vectors which causes node j to be 0. Find the set N_j^1 of input vectors which causes node j to be 1.

Step 2: Repeat step 1 for node k.

Step 3: Compute $T_i = (N_j^0 \cap N_k^1) \cup (N_j^1 \cap N_k^0)$.

Step 4: Use either procedure A or B above to select necessary vectors from step 3.

Few tools list undetected faults in a separate file and also create a file of strobe time for each vector. Cycle splitting has also been used to catch faults during strobes when the clock is high and again when the clock is low. However, this splitting is possible only for return-to-zero (RZ) and return-to-one (RTO) clocks. Generally, one strobe at the end of the cycle is sufficient if only one clock edge latches the data in the storage elements and the other clock edge does not cause any node to switch state. Test patterns for bidirectional pins are such that they change only on the test cycle boundary. When bidirectional pins change state, there is no strobe on one cycle after the enable pin switches its state.

From a user's perspective, the generalized overall flow of an Iddq test tool with necessary file structure is similar to that shown in Figure 9.17. The format of files varies from one tool to another. Also, some tools may not provide all the files as shown in Figure 9.17. The operational mode of tools also varies from interactive mode to batch mode.

A number of commercial tools from EDA companies Sunrise (now Synopsys), Mentor Graphics, System Science (now Synopsys), CrossCheck (now Duet), and Syntest and from IC manufacturers Ford Microelectronics, IBM, Lucent, LSI Logic, Philips Microelectronics, Texas Instruments, and NEC Microelectronics are available for developing an Iddq test set. Some of the tools from IC manufacturers can also be licensed, such as Quietest from Ford Microelectronics, Iddalyzer from LSI Logic, GenTest from Lucent, and Testbench from IBM. Some tools of similar capabilities have also been developed at universities.

There is debate on how many vectors should be used and which are the best vectors. A number of studies including [33] and the Sematech experiment S-121 indicate that the best benefit is obtained from using about 20 vectors obtained for the highest bridging coverage (fault graded under the bridging fault model). Note that these 20 vectors will most likely not provide

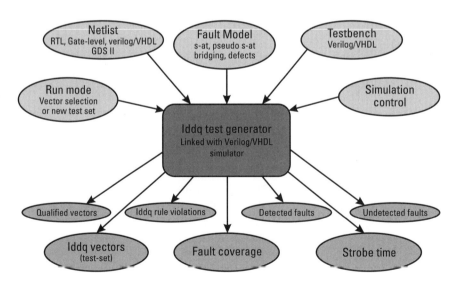

Figure 9.17 Generalized overall flow of Iddq test generator.

100% coverage under any assumed fault model; it is merely a suggested cut-off point on the cost-coverage trade-off curve above which sufficiently more vectors will be required for higher fault coverage. Figure 9.18 shows the nature of coverage by various Iddq test sets.

The fault models used in Iddq test generators are (1) stuck-at, (2) pseudo-stuck-at, (3) toggle coverage, (4) bridging, and (5) defects (in experimental stage). The closest defect-model–based tools are Ford Microelectronics Quietest and CrossCheck's CM-I tools. These tools use a special fault mode for each cell type in which each node in the transistor schematic is analyzed for each vector. This creates a large number of faults even for primitive gates. For example, a two-input NAND gate contains 25 faults in the CrossCheck model. The extremely large fault set makes the simulation slow and, hence, Iddq test generators using this model are slow.

To speed up the process, faults across cell boundaries were considered. It reduces the number of faults, that is, a two-input NAND gate contains only nine faults at the cell boundary and, hence, simulation is faster. Another tool, Power Fault from System Science, can also work at the RTL netlist. Although it is not advisable to develop an Iddq test set using RTL netlist, running the tool at that level identifies Iddq rule violations (such as bus conflicts) in the early stage. Hence, this type of tool also becomes useful in the design process.

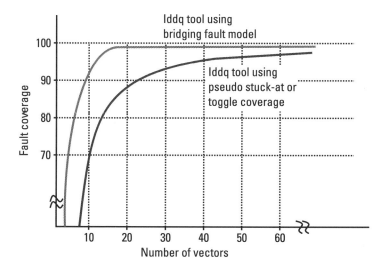

Figure 9.18 Effectiveness of Iddq patterns generated using various fault models.

References

[1]   Keshavarzi, A., K. Roy, and C.F. Hawkins, "Intrinsic leakage in low power deep submicron CMOS ICs," *Proc. IEEE Int. Test Conf.,* 1997, pp. 146–155.

[2]   Rajsuman, R., "Iddq testing for CMOS VLSI," *Proceedings of the IEEE,* vol. 88 (4), April 2000, pp. 1–25.

[3]   Rajsuman, R., *Iddq Testing for CMOS VLSI,* Norwood, MA: Artech House, 1995.

[4]   Maly, W., "Realistic fault modeling for VLSI testing," *IEEE Design Auto. Conf.,* 1987, pp. 173–180.

[5]   Peters, L., "10 significant trends in field emission SEM technology," *R&D Magazine,* Aug. 1994, pp. 52–56.

[6]   Henderson, C. L., and J. M. Soden, "Signature analysis for IC diagnosis and failure analysis," *Proc. IEEE Int. Test Conf.,* 1997, pp. 310–318.

[7]   Xu, S., and S. Y. H. Su, "Detecting I/O and internal feedback bridging faults," *IEEE Trans. Computers,* June 1985, pp. 553–557.

[8]   Malaiya, Y. K., A. P. Jayasumana, and R. Rajsuman, "A detailed examination of bridging faults," *Proc. IEEE Int. Conf. on Computer Design,* 1986, pp. 78–81.

[9]   Zaghloul, M. E., and D. Gobovic, "Fault modeling of physical failures in CMOS VLSI circuits," *IEEE Trans. Circuits and Systems,* Dec. 1990, pp. 1528–1543.

[10] Montanes, R. R., E. M. J. G. Bruls, and J. Figueras, "Bridging defect resistance measurement in a CMOS process," *Proc. IEEE Int. Test Conf.,* 1992, pp. 892–899.

[11] Favalli, M., et al., "Analysis of resistive bridging fault detection in BiCMOS digital ICs," *IEEE Trans. VLSI,* Sep. 1993, pp. 342–355.

[12] Hawkins, C. F., and J. M. Soden, "Reliability and electrical properties of gate-oxide shorts in CMOS ICs," *Proc. IEEE Int. Test Conf.,* 1986, pp. 443–451.

[13] Chan, A., et al., "Electrical failure analysis in high density DRAMs," *IEEE Int. Workshop on Memory Technology, Design and Testing,* 1994, pp. 26–31.

[14] Holland, S., et al., "On physical models for gate oxide breakdown," *IEEE Electron Device Lett.,* Aug. 1984, pp. 302–305.

[15] Bhattacharyya, A., J. D. Reimer, and K. N. Rits, "Breakdown voltage characteristics of thin oxides and their correlation to defects in the oxides as observed by the EBIC technique," *IEEE Electron Device Lett.,* Feb. 1986, pp. 58–60.

[16] Syrzycki, M., "Modeling of gate oxide shorts in MOS transistors," *IEEE Trans. CAD,* Mar. 1989, pp. 193–202.

[17] Montanes, R. R., et al., "Current vs. logic testing of gate oxide short, floating gate and bridging failures in CMOS," *Proc. IEEE Int. Test Conf.,* 1991, pp. 510–519.

[18] Soden, J., and C.F. Hawkins, "Test considerations for gate oxide shorts in CMOS ICs," *IEEE Design and Test,* Aug. 1986, pp. 56–64.

[19] Jacomet, M., and W. Guggenbuhl, "Layout dependent fault analysis and test synthesis for CMOS circuits," *IEEE Trans. CAD,* June 1993, pp. 888–899.

[20] Khare, J., et al., "Key attributes of an SRAM testing strategy required for effective process monitoring," *IEEE Int. Workshop on Memory Testing,* 1993, pp. 84–89.

[21] Moritz, P. S., and L.M. Thorsen, "CMOS circuit testability," *IEEE J. Solid State Circuits,* Apr. 1986, pp. 306–309.

[22] Jha, N. K., "Multiple stuck-open fault detection in CMOS logic circuits," *IEEE Trans. Computers,* Apr. 1988, pp. 426–432.

[23] Reddy, S. M., and M. K. Reddy, "Testable realization for FET stuck-open faults in CMOS combinational circuits," *IEEE Trans. Computers,* Aug. 1986, pp. 742–754.

[24] Liu, D. L., and E. J. McCluskey, "CMOS scan path IC design for stuck-open fault testability," *IEEE J. Solid State Circuits,* Oct. 1987, pp. 880–885.

[25] Jayasumana, A. P., Y. K. Malaiya, and R. Rajsuman, "Design of CMOS circuits for stuck-open fault testability," *IEEE J. Solid State Circuits,* Jan. 1991, pp. 58–61.

[26] Sherlekar, S. D., and P. S. Subramanian, "Conditionally robust two-pattern tests and CMOS design for testability," *IEEE Trans. CAD,* Mar. 1988, pp. 325–332.

[27] Landrault, C., and S. Pravossoudovitch, "Hazard effect on stuck-open fault testability," *Proc. IEEE European Test Conf.,* 1989, pp. 201–207.

[28] David, R., S. Rahal, and J. L. Rainard, "Some relationships between delay testing and stuck-open testing in CMOS circuits," *Proc. European Design Auto Conf.,* 1990, pp. 339–343.

[29] Maly, W., P. K. Nag, and P. Nigh, "Testing oriented analysis of CMOS ICs with Opens," *Proc. IEEE Int. Conf. On Computer Design,* 1988, pp. 344–347.

[30] Soden, J. M., R. K. Treece, M. R. Taylor, and C. F. Hawkins, "CMOS IC stuck-open fault electrical effects and design considerations," *Proc. IEEE Int. Test Conf.,* 1989, pp. 423 430.

[31] Renovell, M., and G. Cambon, "Electrical analysis and modeling of floating gate faults," *IEEE Trans. CAD,* Nov. 1992, pp. 1450–1458.

[32] Champac, V. H., A. Rubio, and J. Figueras, "Electrical model of the floating gate defect in CMOS ICs: implications on Iddq testing," *IEEE Trans. CAD,* Mar. 1994, pp. 359–369.

[33] Maxwell, P. C., et al., "The effectiveness of Iddq, functional and scan tests: how many fault coverages do we need?," *Proc. IEEE Int. Test Conf.,* 1992, pp. 168–177.

[34] Sawada, K., and S. Kayano, "An evaluation of Iddq versus conventional testing for CMOS sea-of-gate ICs," *Proc. IEEE Int. Test Conf.,* 1992, pp. 158–167.

[35] Wiscombe, P. C., "A comparison of stuck-at fault coverage and Iddq testing on defect levels," *Proc. IEEE Int. Test Conf.,* 1993, pp. 293–299.

[36] Chen, C. H., and J. A. Abraham, "High quality tests for switch-level circuits using current and logic test generation algorithms," *Proc. IEEE Int. Test Conf.,* 1991, pp. 615–622.

[37] Storey, T., et al., "Stuck fault and current testing comparison using CMOS chip test," *Proc. IEEE Int. Test Conf.,* 1991, pp. 311–318.

[38] Franco, P., et al., "An experimental chip to evaluate test techniques chip and experimental design," *Proc. IEEE Int. Test Conf.,* 1995, pp. 653–662.

[39] Ma, S. C., P. Franco, and E. J. McCluskey, "An experimental chip to evaluate test techniques experiment results," *Proc. IEEE Int. Test Conf.,* 1995, pp. 663–672.

[40] Nigh, P., et al., "So what is an optimal test mix? A discussion of the SEMATECH method experiment," *Proc. IEEE Int. Test Conf.,* 1997, pp. 1037–1038.

[41] Nigh, P., et al., "An experimental study comparing the relative effectiveness of functional, scan, Iddq and delay-fault testing," *Proc. IEEE VLSI Test Symp.,* 1997, pp. 459–464.

[42] Nigh, P., et al., "Failure analysis of timing and Iddq only failures for the SEMATECH test methods experiment," *Proc. IEEE Int. Test Conf.,* 1998, pp. 43–52.

[43] Ferre, A., and J. Figueras, "Iddq characterization in submicron CMOS," *Proc. IEEE Int. Test Conf.,* 1997, pp. 136–145.

[44] Zarrinfar, F., and R. Rajsuman, "Automated Iddq testing from CAD to manufacturing," *Proc. IEEE Int. Workshop on Iddq Testing,* 1995, pp. 48–51.

[45] Rajsuman, R., "Design-for-Iddq-testing for embedded cores based system-on-a-chip," *Proc. IEEE Int. Workshop on Iddq Testing,* 1998, pp. 69–73.

[46] Rajsuman, R., "Testing a system-on-a-chip with embedded microprocessor," *Proc. IEEE Int. Test Conf.,* 1999, pp. 499–508.

[47] Colwell, M., et al., U.S. Patent No. 5,644,251, July 1997, and U.S. Patent No. 5,670,890, Sep. 1997.

[48] IEEE P1500 Working Group, "Preliminary outline of IEEE P1500's scalable architecture for testing embedded cores," *Proc. IEEE Int. VLSI Test Symp.,* 1999, pp. 483–488.

10

Production Testing

Besides the topics addressed in Chapters 6 to 9, a few other test topics deserve discussion in the context of SoC testing. These topics are related to the production testing of SoC. The objective of production testing is to ensure that the manufacturing is defect-free in a cost-effective manner and also to weed out infant mortality to ensure a specified lifetime.

During production testing, various AC, DC, and other parametric tests are performed multiple times. Because of the cost-quality issues, production testing involves items such as production test flow, at-speed testing, speed binning, test logistics on multiple testers, and production line issues such as multi-DUT testing. We discuss these topics in this chapter.

10.1 Production Test Flow

During the production of ICs, each chip is tested multiple times with electrical AC, DC, and parametric tests. Whereas the packaged part is subjected to all tests, at the wafer level a subset of tests is used. Two examples of production test flow are given in Figure 10.1.

A typical flow after wafer completion may contain the following steps:

1. *Wafer sort testing:* This type of testing contains parametric, functional, analog, and memory tests. Normally the specified junction temperature (T_j) for wafer sort is 75°C. Large memories (such as multimegabit DRAM and flash memories) in SoC are tested

239

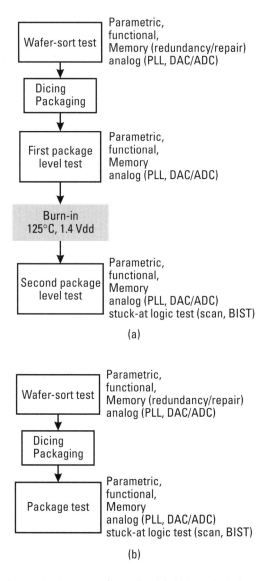

Figure 10.1 Sample production test flows for (a) high-end devices and (b) low-end devices without burn-in.

extensively at this stage with multiple test patterns. All memory failures are identified. This is followed by redundancy analysis and repair. After repair another memory test is performed to ensure that there are no more memory failures.

2. *Dicing and packaging.*

3. *First package level test:* This testing contains parametric and functional tests including a simple test of memory and analog/mixed-signal circuits such as PLL and DACs. The main objective of this test is to filter failures that occurred during packaging.

4. *Burn-in:* Anywhere from a few hours (say, 8 hours) to 24 hours are used for burn-in. Typically, a burn-in temperature of 125°C and a 40% higher voltage (1.4 times Vdd) is used. Note that all manufacturers do not perform burn-in. Also, a few manufacturers have used Iddq or stressed-Iddq testing (40% voltage stress) to either reduce or completely eliminate this burn-in step from their production flow [1].

5. *Second package level test:* This is considered the most important test. It contains all parametric, functional, stuck-at logic (scan, BIST, and so on), full memory and analog/mixed-signal, and Iddq testing. Many manufacturers also apply functional, stuck-at logic, full memory, and analog/mixed-signal tests at more than one voltage level, such as at typical Vdd and at 15% higher voltage (1.15Vdd).

10.2   At-Speed Testing

At-speed testing refers to the execution of functional vectors at the chip's specified speed at which it will operate in the system. At-speed testing is very desirable to detect resistive vias and contacts, partial opens, metal cracks, high-resistance bridges, and timing-related failures. It also provides the maximum confidence that the chip will work in the system.

10.2.1   RTD and Dead Cycles

Implementation of at-speed testing is quite complicated due to the ever-increasing operating speed of the SoC. At chip operating speeds above 250 MHz, round-trip delay (RTD) from the tester driver to the DUT and back to the tester comparator plays a major role in the test program development for at-speed testing. As illustrated in Figure 10.2, the path of RTD contains the device pin, socket pin, trace on the load board, and the wire in the test head to/from the comparator/driver [2]. In general, RTD is more than 2 ns on the present ATE systems.

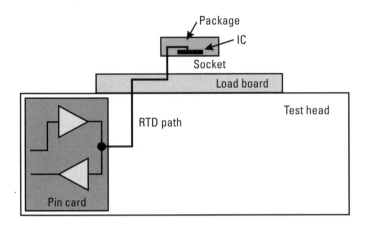

Figure 10.2 Illustration of RTD path from tester driver to DUT and back to tester comparator.

The RTD time puts a restriction on the speed at which a tester can apply a test vector to a bidirectional pin. The basic issue is that the path between the device pin and the tester is shared between the tester driver and comparator. The drive signal from the tester to the DUT must therefore wait until the DUT data has reached the comparator and the tri-state condition has had enough time to propagate through this one-way path. Thus, to test a bidirectional pin, the switching speed from the input state to the output state should be less than the physical delay over this path. Figure 10.2 illustrates a typical bidirectional signal. The maximum frequency at which tester can test is defined by:

Maximum test frequency = Setup time (A) + Active to tri-state interval time (G) + Data-to-tri-state-transition propagation time until it reaches to the comparator (H) + Time for driver signal to reach DUT (I).

Various time segments are illustrated in Figure 10.3. Here, H and I each equal to ½ RTD. Assuming 1-ns RTD and 1 ns for setup (A), and 1 ns for active to tri-state time (G), a total delay of 3 ns; the maximum test frequency can be about 334 MHz (1 tester period = 3 ns) even if the tester runs at higher speed such as 500 MHz.

To overcome this problem during production testing, a concept known as *dead cycle* is used. Dead cycles are no operation (NOP) instructions inserted into the device simulation every time a device pin changes from the

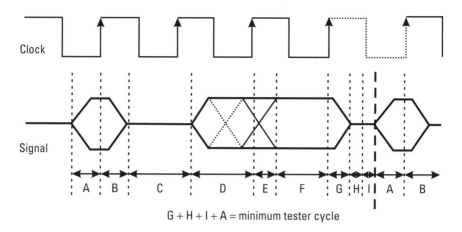

Clock

Signal

A : B : C : D : E : F : G :H: I : A : B

G + H + I + A = minimum tester cycle

Figure 10.3 Illustration of a typical bidirectional tester cycle.

output mode to the input mode. These NOPs provide extra delay to match RTD during switching while the rest of the testing is done at speed. Using the same numbers as before (RTD = 1 ns, A = 1 ns, and G = 1 ns), with one dead cycle, the maximum test frequency can be about 667 MHz (2 tester periods = 3 ns). Thus, the insertion of dead cycles helps; however, certain issues arise when using dead cycles. Chief among these is the fact that due to the presence of a dead cycle, the testing no longer uses the same stimulus it used in the final application, and hence the purpose of at-speed testing is defeated.

10.2.2 Fly-By

To overcome the limitation of dead cycles, another concept known as *fly-by* is used. In fly-by methodology, one tester pin drives data to the DUT, while another tester pin is used to receive and compare. The shared one-way path is no longer used. This provides a one-way path from the tester driver to the DUT pin as well as one from the DUT pin to the tester comparator. In principle, it is simple; however it requires the following tasks:

1. Driver and comparator lines are separate but connecting to the same DUT pin. When each line is terminated at 50Ω, the DUT output sees a 25Ω transmission line (parallel connection of two 50Ω lines), whereas the tester driver sees a 50Ω transmission line. Because the DUT output sees a 25Ω transmission line, it results in

a 50% reduction in the signal amplitude at the tester comparator. This problem can be solved by proper impedance matching.

2. Pattern data must be created that treats the device pin as two separate pins (input and output).

3. Pattern data should still tri-state tester driver for read and compare cycles.

4. Due to high-speed testing and line termination not being 50Ω, signal reflections result on the line. Because of the separation of driver and comparator lines, it is very difficult to perform time-domain reflectometry (TDR) measurement.

All of these tasks require a significant amount of manual work during test program development. They also require twice the number of tester channels because the driver and comparator are separate. With the maximum number of tester channels being 1024 in the present-day ATE systems, this method does not allow at-speed testing of a device with more than 512 pins.

To resolve this issue and to support fly-by, ATE vendors have developed the pin electronics architecture illustrated in Figure 10.4. This architecture allows the use of a tester channel as an input (I), an output (O), a bidirectional pin (I/O), split input and output (I and O) for two different device pins, or split input and output (I and O) for fly-by on one device pin [3]. As illustrated in Figure 10.4, a multipin per channel pin electronics card offers one driver and two comparators interfacing with the device through two high-speed pins.

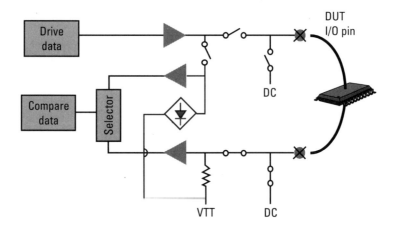

Figure 10.4 Fly-by architecture of tester pin electronics.

In production testing, this is quite beneficial because the tester can be used in the following manner:

1. *Pattern multiplexed mode:* In this mode, two vectors from pattern memory can be applied to the device during each tester cycle. This effectively doubles the programmed pattern rate without the loss of a pin channel. Thus, a tester with a base rate of 250 MHz can test up to 500 MHz.

2. *Pin multiplexed mode:* In this mode, pattern data from two adjacent pin channels can be applied to the device during each tester cycle. This effectively doubles the programmed pattern rate without the loss of pin channel, and a tester with base rate of 250 MHz can test up to 500 MHz.

3. *Combined pin-pattern multiplexed mode:* This combines pin and pattern multiplexing and thus effectively quadruples the programmed pattern rate. In this mode, all tester channels operate in pattern-multiplexed mode, and pin-multiplex can be selected on a per DUT pin, per tester channel basis. Thus, a tester with a base rate of 250 MHz can test up to 1 GHz.

10.2.3 Speed Binning

Speed binning is a sorting process to identify which part will work at what speed. The major issue in speed binning is how to set up the bins and the test flow. Because speed binning requires some parts to be tested multiple times, setting a careful test flow is essential for the production-line throughput.

A postpackage speed binning flow based on higher frequency to lower frequency for the sorting of parts at two frequencies (350 and 300 MHz) is illustrated in Figure 10.5(a). In this flow, parts in bin A (300 MHz) and parts that will be scrapped (bin C) are tested twice. An alternate flow based on lower frequency to higher frequency is illustrated in Figure 10.5(b), in which parts are first tested at 300 MHz and then at 350 MHz. In Figure 10.5(b), parts in bin A and bin B are tested twice.

On the surface, it is difficult to say which flow is better. However, if the production yield is high, the first flow [Fig. 10.5(a)] should be chosen. If the production yield is low, the second flow [Fig. 10.5(b)] is preferable. To minimize the total test time, smart test flows for speed binning have also been developed that switch from lot to lot based on wafer yield from one setup to another.

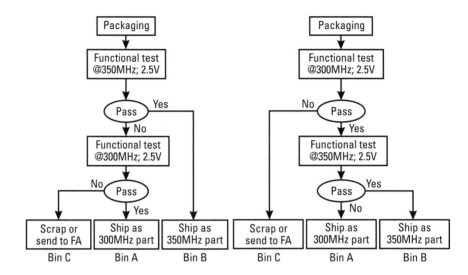

Figure 10.5 Simple postpackaged speed bin flow for two frequencies: (a) high to low frequency and (b) low to high frequency.

10.3   Production Throughput and Material Handling

Similar to any factory or manufacturing process, the key issue in material handling is the throughput of the production line. In SoC testing, the test flow and setup as discussed in Sections 10.1 and 10.2.3 can significantly impact test time per chip and subsequently tester throughput and the test cost. Other critical factors for production throughput are test logistics, tester setup time, and the handler index.

10.3.1   Test Logistics

Test logistics and selection of the right tester platform are extremely important in SoC testing. The key issue is how to test large embedded memory. Embedded memories in SoC require algorithmic pattern generation, redundancy analysis, and repair capabilities. A small number of tester pins is sufficient for memory testing, and at the same time it is highly desirable to test multiple chips in parallel. Thus, memory testers are ideally suited because logic testers are high pin-count testers with limited algorithmic pattern generation (ALPG) capability. If a memory ALPG unit is integrated with a logic tester to test embedded memories, a large number of pins of the logic tester remain idle during memory testing and hence the utilization of tester resources remains poor. If, on the other hand, embedded memory is tested on a memory tester, complete SoC testing requires two setups with loading/unloading on two testers (dual insertion on two testers). The overhead time thus becomes significant and, depending on the amount of testing, it may or may not be cost effective.

To provide a guideline and to quantify costs to determine when to use a memory tester to test embedded memory, a model was described in Section 7.5. To reiterate, dual insertion is generally cost effective for testing 32 or more embedded memories that are 16 Mbits or larger.

A very similar situation arises for testing analog/mixed-signal circuits such as DACs and PLLs. Fortunately, most ATE vendors now provide sufficient analog/mixed-signal test capability with the logic tester that these circuits can be adequately tested in a cost-effective manner on the logic tester.

10.3.2   Tester Setup

Another item that affects SoC production throughput is the tester setup time. From a throughput point of view, it is pure overhead. Generally, four or five designs (sometimes even more) are assigned to one tester platform. Tester time is thus scheduled on a batch-by-batch basis—for example, 8 a.m. to 9 a.m., testing of design A; 9 a.m. to 12 noon, design B, and so on. Each design requires its own test setup and test patterns. Hence, when a batch is changed, the tester requires a new setup and test pattern. For production line throughput, this setup and pattern loading time needs to be as short as possible.

The size of SoC test patterns is considerably large (on the order of millions of vectors), so to transfer this data to a tester controller and to then load it onto a pattern generator (PG) would take a significant amount of time. For example, an SoC with 16 logic pattern (LPAT) files—each file containing 640K vectors—implies approximately 10.24M vectors. This is an enormous amount of data for a 512-pin SoC (each vector being 512 bits long). For this example, the pattern load time from tester controller to pattern generator (PG) will take approximately 34.5 minutes on a commercial ATE. The tester cannot perform testing during this time and, hence, it is pure overhead.

The solution to this problem is to develop a mechanism for high-speed pattern loading. Some ATE companies have developed a high-speed pattern loading server. Such a server contains a large amount of pattern storage (on the order of 100 GB) that is used to store multiple test setups for one or multiple testers. In principle, such servers behave in a manner similar to a computer server. For example, a high-speed pattern loading server from

Advantest (W4322) provides 72 GB of pattern storage (expandable to 504 GB) and supports up to eight test systems. Using the number in our previous example (16 LPAT files, each with 640K vectors, a total of 10.24M vectors), this server will take approximately 83.5 sec to load these vectors to PG. From a production line point of view, it reduces tester setup overhead time from approximately 34.5 minutes to 1.5 minutes.

10.3.3 Multi-DUT Testing

Another desirable item from the production throughput point of view is the testing of multiple chips at the same time. Two methods are available to test multiple chips at the same time:

1. *Multi-DUT load board:* In this approach, the load board is designed with multiple sockets to support loading/testing in parallel. The test input is basically fanout to all the sockets and, hence, all chips get the same test stimulus. For response evaluation, the tester pins are grouped in such a way that each group evaluates response from one socket. The schematic of this load-board structure is illustrated in Figure 10.6. Using this method, memory manufacturers have tested as many as 64 memory chips in parallel. In logic testing, however, this method is limited to 2 to 4 chips, depending on the number of signal pins and tester configuration. In SoC testing, this

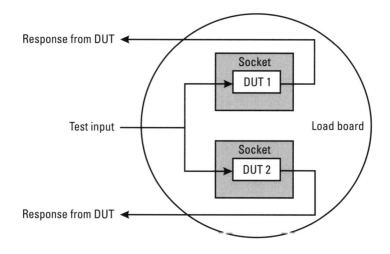

Figure 10.6 Illustration of dual-DUT load board.

method is particularly useful for testing large embedded memories on the memory tester.

One little complication to this method is the fact that the test continues until the end, even if one chip fails. Also, the loading and unloading of multiple chips and keeping track of failed chips requires a little care in the test setup. Due to these factors, the tester throughput does not increase by n-fold if n chips are tested in parallel. As a general guideline, tester throughput increases by a factor of 1.6 to 1.8 for dual-DUT and 2.7 to 3.6 for quad-DUT testing.

2. *ATE with multiple test heads:* In this approach, multiple test heads (generally dual test heads) are used. The test program and test setup are the same for both test heads. The test input signal in this case is fanned out inside the tester pin electronics rather than at the load-board level. Note that multi-DUT load boards could also be used with each test head to obtain the maximum parallelism.

References

[1] Rajsuman, R., "Iddq testing for CMOS VLSI," *Proceedings of the IEEE,* vol. 88 (4), April 2000, pp. 1–25.

[2] Katz, J., "High speed testing—have the laws of physics finally caught up with us?" *Proc. IEEE Int. Test Conf.,* 1998, pp. 809–813.

[3] "T6682 pin and pattern multiplex operation for high speed functional test requirements," Application Note AN0025A, Advantest Corp., 1998.

11

Summary and Conclusions

This book has attempted to capture most of the necessary and fundamental topics related to SoC design and testing. In the first part (Chapters 1 to 5), an overview of design methodologies for SoC was given, whereas the second part (Chapters 6 to 10) covered SoC testing.

At the present time there are a number of open design and test issues that are not well understood by the design/test community because they have very little to no support from the EDA tools point of view (these issues were discussed in the Chapter 1). Some key items from Chapters 1 to 10 are summarized below, followed by a couple of possible future scenarios.

11.1 Summary

It is clear that at the present time system-on-a-chip is not just a large ASIC; it is a system that requires design efforts from both the hardware and software sides. Thus, hardware–software codesign is essential for SoC.

Core-based design reuse is necessary to manage the design complexity of SoC. Design-for-reuse requires a strict set of rules both at the core and at the SoC level. These rules needs to be observed at the design specification level and should be followed through to the individual core level and finally at the SoC level.

The design, delivery, and productization of cores requires multiple HDL models, multiple functional and timing simulation models, complete testbenches, and detailed documentation. The soft cores provide maximum flexibility in the design and productization, but also result in the least predictability in the areas of area and performance. Hard cores provide reasonably predictable performance and area, but they have strict design and productization requirements.

A restricted set of RTL rules is necessary in core design to facilitate portability and reusability. In addition, RTL should be geared toward deterministic synthesis; the RTL code should not contain elements that are subjected to open interpretation by the synthesis tool. Furthermore, RTL should be checked and made free of Lint-based errors and warnings.

The integration of large DRAM and flash memories presents unique design and process issues. Integration of DRAM into a logic process causes high leakage or requires multitransistor DRAM cells; on the DRAM process, logic performance is limited. For optimized DRAM and logic, a dual-gate process is required. Similarly, the integration of flash memory requires a dual-gate and dual-poly process.

The memory compilers optimized for specific fabrication processes will become increasingly available for SRAM, DRAM, ROM, EEPROM, and flash memories. The foundry companies will also provide a large number of custom and semicustom memory cores as well as analog/mixed-signal cores such as PLL and DAC/ADC that are optimized for their processes.

System validation is one of the most difficult tasks in SoC development. A detailed validation plan and testbenches are required to validate (1) core functionality and timing, (2) core interfaces, and (3) SoC-level functionality and timing.

From the beginning of the design phase, the SoC validation strategy should be defined. The periodic regression testing should contain directed tests as well as random tests. One or more directed tests should be added to the regression test suite for every new bug. A careful data log on bugs (date and how they are found) should be provided by an automatic bug-tracking system. Such a data log system also helps in understanding the stability of the design.

SoC-level verification should be done in a combined hardware–software coenvironment and should contain multiple application runs. It should be done through hardware–software cosimulation, through emulation, or through direct runs on hardware prototypes.

Core testing can be facilitated by a boundary-scan-type wrapper to address test access, isolation, test control, and observation. This wrapper can be either integrated into the design itself or can be made into a separate file. The testing of desktop designs will be highly dependent on the wrapper. The ASIC vendor designs can merge core wrappers into the design for

SoC-level optimization. With the wrapper, robust functional verification and test vectors can be developed.

Hierarchical TAP controllers can be used in desktop designs. EJTAG and similar modifications of the IEEE 1149.1 JTAG standard are very useful to support core and SoC testing, debugging, characterization, and production testing.

Detailed debugging features need to be implemented in the cores as well as in the SoC. The availability of debuggers in the open market is expected to increase dramatically once EDA companies have established themselves as IP brokers. However, at the present time it is a bottleneck because individual debuggers are problematic in addressing unique data transactions among multiple cores in SoC.

Full scan is the primary test structure internal to the digital logic core. The use of logic BIST based on simple LFSR and MISR in data path circuits will increase; however, the use of logic BIST structures such as scan-based BIST or arithmetic BIST is very limited.

High-level test synthesis is an active research topic with increasing focus on DSP cores. However, without a breakthrough from EDA companies in developing a robust tool, there is very limited use of high-level test synthesis in the core test implementation.

In the high-end applications, the on-line testing of SoC is quite useful. Many of the on-line test functions are based on self-testing, using one core to test other cores. This behavior will further expand as testing parts of the system in watchdog mode during normal operation of the system.

The use of memory BIST with a shared controller is essential. The memory BIST with diagnostic features to support BISR is necessary for large embedded memories. The memory BIST controllers will also evolve into hierarchical form to test memories embedded in a core as well as memories at the SoC level. These controllers need to be enhanced so that they can also provide failed bit maps in the diagnostic mode.

With large embedded DRAM and flash, more and more ASIC vendors and foundries are using dual insertion. The memory BISR functions are carried out on memory testers. Besides memory BISR, another class of BISR functionality will be used to disable a faulty digital logic core and provide that functionality in the software. Another possibility is the reconfiguration of SoC into a product with subfunctionality.

A new class of fault behaviors will arise on SoC with large embedded DRAM and flash memories that are fabricated in a process optimized for digital circuits. Similar situations will arise when SoC is fabricated in a process optimized for DRAM.

Embedded mixed-signal circuits will continue to be tested in the production line for a few key parameters. The use of analog BIST as well as 1149.4 is very limited.

Iddq testing will continue to be useful in the production line as a supplemental test. The primary method will be the performance of Iddq testing on one core at a time. SoC-level Iddq testing will be based on consolidation of Iddq vectors of the individual cores. In very few cases, additional chip-level Iddq vectors will be developed.

Timing-based testing will be increasingly used as part of the characterization as well as production test. Selection of a few paths, on-chip monitors for delay variations, delay fault testing, and transition-propagation-based testing will be used increasingly in the characterization phase and in a limited form during production testing.

At-speed testing, speed binning, smart test flow, and material handling under multiple insertions are serious issues for foundries, ASIC vendors, and fabless semiconductor vendors.

Large amounts of data and tester overhead are a key issue in the production line throughput.

11.2   Future Scenarios

Based on trends in technology progression, a couple of possible future scenarios (for the approximately 2005 time frame) for SoC design and testing and overall industry are discussed next.

Both ASIC vendors and design houses will produce more and more integrated designs; design houses and the companies providing design services will gear more toward desktop designs (the definitions of integrated and desktop designs are given in Chapter 1). A new breed of company may also emerge that will sell chips but will essentially be IC specs developers that become designless-fabless semiconductor companies.

These companies will utilize IP brokers, design services, and semiconductor foundries to design and fabricate the chips. The process technology and maximum operating speed of such chips will lag one to two generations behind the high-end microprocessors as well as ASIC vendor designs.

Most of the EDA companies will also serve as IP brokers and will provide certified cores on a select set of tools. Most of this data as well as EDA tools will be available through password-protected Internet access. Besides design, Internet hook-ups will also facilitate SoC testing for fabless companies.

Design verification and validation will be the key issues in integrated and desktop designs and may hinder progress toward Internet-based design/test processes. Design verification may also become a limiting factor in the viability of designless-fabless semiconductor companies.

Appendix:
RTL Guidelines for Design Reuse

The basic requirement of design reusability is that the RTL code be portable. For hard cores, good RTL code is also necessary to obtain good netlist during synthesis. RTL coding is equivalent to software development; hence, commonly used software-engineering guidelines are applicable during RTL coding.

Similar to software development, the basic coding guidelines require that the RTL code be simple, structured, and regular. Such code is also easy to synthesize and verify. In this appendix, we give some basic guidelines. Note that these guidelines should be used only as a reference; each design team should develop their own design guidelines tailored to the design environment, tools, and specifications of the product.

A.1 Naming Convention

One of the main difficulties in design reuse is the lack of a consistent naming convention. A consistent and suitable naming convention should be developed at the beginning of the design phase and should be enforced by the entire design team. Simple naming guidelines include the following:

1. Use lowercase letters for all signals, ports, and variable names. Use uppercase letters for constant names, user-defined types, and parameters.

2. Names should be meaningful and should explain the behavior of the variable. For example, use "rst," "clr" or "reset," "clear" for reset signals. These signals should also follow the active low/high conventions.

3. Parameter names should be kept short. Because synthesis tools concatenate names during processing (elaboration) to create unique parameter names, long names cause readability problems.

4. Use "clk" or "Clk" to name clock lines, or use it as a prefix for all clock signals, for example, clk_slow, clk_A. Also, use the same name for all clock signals that originated from one source.

5. Differentiate between active low and active high signals. One simple method is to use "_h" to indicate active high signals, and "_l" for active low signals.

6. When declaring multibit variables, use a consistent order of bits, either high to low or low to high—that is, 7 down to 0 for VHDL, or (7:0) for Verilog. If zero is used as the low-end bit, then all multibit variables should be in the form of $(n-1)$ to 0. These consistencies avoid coding bugs when multibit variables are connected.

7. Use a meaningful label name for process blocks. Again, label names should not be long. Do not duplicate names; for example, giving the same name to a process as well as a signal will result in confusion.

If the design is done in VHDL, remember that IEEE Standard 1076.4 provides certain restrictions regarding naming conventions of port declarations at the top level of library cells. Particularly, level 0 compliance has rules for naming port/generic types so that simulator can handle SDF back-annotations in a uniform way.

A.2 General Coding Guidelines

1. Use simple constructs and clocking schemes. The structure of the code should be consistent. Consistency can be achieved by a regular partitioning scheme that makes module sizes approximately the

same. Partitioning of the design should be such that all logic in a single module uses a single clock or a reset.

2. The coding style should be consistent and based on the usage of parameters rather than constant numbers.

3. Indentation should be used to improve the readability of the RTL code. However, it should be small (2 to 3 spaces), to keep the line length small. The [Tab] key should be avoided because tabs can misalign text in different text editors.

4. Port declarations should be logical and consistent, that is, all inputs, clocks, outputs, and inouts should be in the proper order. It is useful to declare one port per line with a short comment describing the port. When a module is instantiated, declaration should also be explicitly mapped using names rather than positional association.

5. Use functions wherever possible instead of repeating the code. It is a good practice to generalize the function for wider reuse.

6. Loops and arrays should be used for code compaction. Arrays are faster to simulate than loops and, hence, simulation performance can be improved by using vector operations instead of loops.

7. Internal three-state buses present serious design challenges and should only be used when absolutely necessary.

8. Coding should be technology independent and compatible with various simulators. In VHDL, only IEEE standard types should be used.

9. The translation of VHDL designs into Verilog is often necessary for various reasons. Some VHDL constructs such as block constructs generate statements that modify constant declarations and have no equivalence in Verilog. These should be avoided in RTL.

10. During the development of cell libraries, emphasis should be on technology independence. Hence, avoid instantiating gates in RTL code. Gate-level designs are hard to read and hence hard to maintain and reuse.

11. In high-speed circuits, data needs to be captured on both edges of the clock pulse. Thus, in high-speed circuits, the duty cycle of the clock is critical for timing analysis. Synthesis and timing tools should get the worst case duty cycle of the clocks and that duty cycle should be clearly documented. If mixed clock edges are necessary, then they should be grouped into separate modules.

12. Avoid internally generated clocks. Similarly, internally generated conditional resets should be avoided. Internally generated clocks and resets are problematic in testing and they make synthesis difficult. The exception to this rule is low-power designs that need gated clocks; otherwise, gated clocks should be avoided. In this case, clock-gating logic must be kept at the top level in a separate module.

A.3 RTL Development for Synthesis

The objective of RTL is the creation of design through synthesis process. Every synthesis tool has its own flavors for HDL constructs; these constructs make the synthesis process efficient and also simplify postsynthesis analysis. In addition to the general coding guidelines as given in Section A.2, RTL development should be targeted toward synthesis. Some generalized guidelines for making RTL synthesis friendly are given as follows:

1. Latches should be avoided in the design whenever possible. The synthesis tool can create latches due to incomplete or ambiguous RTL code. Hence, when latches are necessary, it is cleaner to instantiate them rather than let the synthesis tool introduce them in the netlist.

2. Another important item to avoid in RTL is feedback combinational loops. Feedback combinational loops create various timing issues in the design reuse. Similarly, asynchronous logic should be avoided whenever possible.

3. Registers and flip-flops provide synchronization and are hence preferred over latches for sequential logic. Instead of using an "initial" statement (Verilog) or "initialization" at the declaration (VHDL), a circuit reset signal should be used to initialize the registers. Without explicit use of reset, the intended reset mechanism is not guaranteed in the netlist. For example, in Verilog, the first "always" block is a register with synchronous reset, whereas the second "always" block is an asynchronous reset.

4. The sensitivity list should be as complete as possible. When sensitivity lists are incomplete, simulation mismatches between pre- and postsynthesis netlists are likely. In combinational blocks, the sensitivity list must contain every signal read in by that process/block.

For sequential blocks, the sensitivity list must include clock and other control signals. Unnecessary signals in the sensitivity list should also be avoided, because they slow down the simulation.

5. In Verilog, nonblocking assignments should be used for clocked blocks. Otherwise the RTL- and gate-level behavior may differ. Similarly, in VHDL, it is better to use signals instead of variables.

6. Case statements in both Verilog and VHDL represent a single multiplexer, while if-then-else statements create a cascaded multiplexer chain. It is thus better to avoid if-then-else statements that have more than two conditions.

7. The RTL code should be partitioned in certain ways so that the synthesis process works efficiently and timing requirements can be tried out easily. As an example, when a state machine is coded, the code can be separated into two processes—one for the combinational logic and another for the sequential logic.

8. Whenever possible, the output of a block should be registered. This simplifies synthesis and makes output drive strength and delays predictable. It also makes technology mapping easier and extremely useful in the design reuse.

9. Multicycle paths should be avoided for efficient timing analysis. If a multicycle path is absolutely necessary, then it should be contained within a single module so that the path does not cross module boundaries.

10. Related combinational logic should be kept in the same module. Modules with different synthesis goals should be separated. Grouping the related logic functions is also better to avoid interblock timing dependencies.

11. Instantiation of gate-level logic at the top level of the design should be avoided. Logic at the top level prohibits tools to optimize this logic with the block's logic. If multiple copies of a core are used in SoC, then glue logic should be placed inside the core. For example, clock domain buffers or bus interface gates can be placed inside the cores so that the glue logic becomes simple interconnecting wires.

12. Synthesis directives (such as "dc_shell" commands) should not be embedded in the RTL. If synthesis goals are different from one SoC to another, the embedded commands will cause conflicts and will limit design reuse. One exception to this rule is the inclusion of directives that turn the synthesis process on and off.

A.4 RTL Checks

Synthesis tools interpret the RTL code in certain ways to decide on what type of logic it implies. Since the synthesis process itself takes a considerable amount of time, there are tools that can browse through the RTL and point out potential problems and errors during the synthesis. Such tools are called *Lint tools,* named after software programming tools such as Glint or Clint. One widely used Lint tool is Verilint (from interHDL) for Verilog.

It is highly recommended that designers integrate a Lint filtering process into the design flow such as in the makefile. This process will detect many design errors and code style violations early on and force engineers to check in only verified code.

Some conditions and situations for which Verilint will give warnings are as follows:

1. When all possible conditions are not given in a Verilog case statement and a "full_case" directive is specified for the synthesis tool.

2. When a net is driven by multiple drivers but is not declared to be a tri-state net.

3. Any operand to a logical binary operation (for example &&) has more than one bit. This is only a warning since the tool can infer default logic gates for such a case.

4. When a self-assignment is present in the code (such as a = a), since it will create a latch.

5. Constant in the sensitivity list because it has no effect.

6. If a loop index variable is modified inside the body of the loop.

7. When a variable is referenced in a nonblocking assignment that was previously assigned but not present in the sensitivity list of the "always" block. If the statements are blocking assignments, it is fine since the variable acts as a temporary variable.

8. If a variable is assigned in both blocking and nonblocking assignments. Tools cannot decide whether it is a register/latch or a temporary variable.

9. If different bits of a variable are assigned values in multiple "always" blocks.

10. When a function is being assigned using the "assign" construct rather than "=", since the semantics are not clearly defined in Verilog.

11. When complex condition expressions are used that could be interpreted differently due to precedence rules.

12. When blocking or nonblocking assignments have delays associated with them, since synthesis does not derive delays from the RTL statements.

13. When RTL constructs are mixed inside gate-level designs. Having few gates inside RTL is not a warning. Also, continuous assign statements are allowed inside gate-level blocks.

14. When a variable is set but not used inside the RTL code.

15. Some compiler constructs such as "ifdef" or "ifndef" may not be supported by synthesis tools—hence, Verilint gives warning.

16. When input or an inout is declared as a registered variable.

17. When the "or" operator is used in a potentially illegal way such as "a & b or c."

Some conditions and situations for which Verilint will issue errors are as follows:

1. When an array is accessed by a negative index.

2. When an instance name (such as of a UDP) is in a sensitivity list.

3. An operand to a logical NOT operation has more than one bit since a logic gate cannot be inferred correctly by the tool.

4. When integer and constant case expressions are out of range.

5. Some synthesis tools do not allow bit selection in the sensitivity lists and, hence, Verilint gives an error message. Also, arrays cannot be used in the sensitivity lists.

6. When both leading and trailing edges are used in an "always" block.

7. When global variables are used inside functions or tasks.

8. All syntax errors and semantic errors in Verilog are also flagged as errors by Verilint.

About the Author

After receiving his Ph.D. in electrical engineering from Colorado State University, Rochit Rajsuman served on faculty in the department of Computer Engineering and Science at Case Western Reserve University for almost seven years. He later left academia to join LSI Logic as a product manager for test methodologies, where he productized a number of test solutions, including Iddq testing. From LSI Logic, Dr. Rajsuman moved to a media processor start-up. He now works as Manager of Test Research at Advantest America R&D Center.

He has authored a number of patents and has published over 60 papers in refereed journals and conferences. He has authored two books—*Digital Hardware Testing* (1992) and *Iddq Testing for CMOS VLSI* (1995), both published by Artech House. He also co-edited the first book on Iddq testing, *Bridging Faults and Iddq Testing* (IEEE Computer Society Press, 1992).

In 1995, Dr. Rajsuman co-founded the *IEEE Int. Workshop on Iddq*. He also co-founded two other international conferences—the *IEEE Int. Workshop on Memory Technology, Design, and Testing*, and the *IEEE Int. Workshop on Testing Embedded Core-based Systems*. He now serves on the Steering Committee of all three workshops. Dr. Rajsuman was also a founding member of the IEEE P1500 Working Group to define embedded core test standards. He serves on the technical program committees of numerous other conferences, including the *International Test Conference*. He is a senior member of IEEE, a Golden Core member of the IEEE Computer Society, and a member of both Tau Beta Pi and Eta Kappa Nu.

Index

Related Titles from Artech House

Electrical and Thermal Characterization of MESFETS, HEMTs, and HBTs, Robert Anholt

An Introduction to Micromechanical Systems Engineering, Nadim Maluf

High-Level Test Synthesis of Digital VLSI Circuits, Mike Tien-Chien Lee

Iddq Testing for CMOS VLSI, Rochit Rajsuman

Principles and Analysis of AlGaAs/GaAs Heterojunction Biploar Transistors, Juin J. Liou

Reliability and Degradation of III-V Optical Devices, Osamu Ueda

System-on-a-Chip: Design and Test, Rochit Rajsuman

For further information on these and other Artech House titles, including previously considered out-of-print books now available through our In-Print-Forever® (IPF®) program, contact:

Artech House
685 Canton Street
Norwood, MA 02062
Phone: 781-769-9750
Fax: 781-769-6334
e-mail: artech@artechhouse.com

Artech House
46 Gillingham Street
London SW1V 1AH UK
Phone: +44 (0)20 7596-8750
Fax: +44 (0)20 7630 0166
e-mail: artech-uk@artechhouse.com

Find us on the World Wide Web at:
www.artechhouse.com